# Resigned Activism

**Urban and Industrial Environments**

Series editor: Robert Gottlieb, Henry R. Luce Professor of Urban and Environmental Policy, Occidental College

*For a complete list of books published in this series, please see the back of the book.*

# Resigned Activism

## Living with Pollution in Rural China

Anna Lora-Wainwright

The MIT Press
Cambridge, Massachusetts
London, England

This book was set in ITC Stone Sans Std and ITC Stone Serif Std by Toppan Best-set Premedia Limited. Printed and bound in the United States of America.

Library of Congress Cataloging-in-Publication Data

Names: Lora-Wainwright, Anna, 1979– author.
Title: Resigned activism : living with pollution in rural China /
    Anna Lora-Wainwright.
Description: Cambridge, MA : The MIT Press, 2017. | Series: Urban and industrial
    environments | Includes bibliographical references and index.
Identifiers: LCCN 2016040529| ISBN 9780262036320 (hardcover : alk. paper) | ISBN
    9780262533850 (pbk. : alk. paper)
Subjects: LCSH: Pollution–Health aspects–China. | Environmental policy–China–
    Citizen participation. | Rural health–China. | Rural development–Environmental
    aspects–China. | China–Rural conditions. | China–Environmental conditions.
Classification: LCC TD187.5.C6 L67 2017 | DDC 363.730951/091734–dc23 LC
    record available at https://lccn.loc.gov/2016040529

10  9  8  7  6  5  4  3  2  1

*A Leon, Robin, e Luca*

# Contents

# List of Figures

All photographs were taken by the author.

# Acknowledgments

Pollution appears to be a persistent and pervasive ingredient of many of our lives. In 2016, my native province, Vicenza (Italy), was the subject of intense debate after a study revealed sixty thousand people were "poisoned" by polluted water. London, where I currently live, routinely makes the news for exceeding EU limits on air pollution. News items on pollution in China are almost too frequent to follow. But, of course, pollution does not touch all of us in the same way. Doing research on such a complex and sensitive topic in China required time, patience, and perseverance. There were times when I wondered whether this book would ever materialize. That it did is only thanks to all the support I have received from so many individuals and institutions.

The British Interuniversity China Centre funded a research fellowship at the University of Manchester (2007–2009) during which I first developed the ideas and networks underpinning this book. The Contemporary China Studies Program at the University of Oxford covered the cost of my salary (2009–2011) while I continued to undertake fieldwork and began preliminary data analysis. A grant by the China Environment and Health Initiative (CEHI) at the Social Science Research Council (SSRC) for the project *Citizens' Perceptions of Rural Industrial Pollution and Its Effects on Health* (RBF/SSRC-CEHI/2008–01–07) supported my fieldwork in Baocun. A British Academy Small Grant for the project *Making Environmental Health Subjects in Contemporary Rural China* (SG091048) funded further fieldwork in Baocun and in Qiancun. The John Fell OUP research fund at Oxford University for the project *Urban Mining, Toxic Payload* funded research in Guiyu during 2012–2013. A Leverhulme Trust Research Fellowship (RF-2012–260, 2012–2013) for the project *Living with Pollution in China: An Ethnographic Perspective* allowed me to take a year of research leave, which was spent in

China combining academic exchange and fieldwork. The Victor and William Fung Foundation supported Luo Yajuan's visit to Oxford in the summer of 2013 during which she capably helped me survey Chinese-language literature on cancer villages. The University of Oxford supported a term of sabbatical in 2013 during which I continued data analysis and writing. Oxford's Contemporary China Studies Program and the *Environmental Cultures Network* (funded by AHRC, ESRC, and HEFCE under the British Interuniversity China Centre phase two) supported two separate events (in 2011 and 2013, respectively), which helped me considerably in advancing my thinking. The Philip Leverhume Prize in Geography, which I was awarded in 2013, funded a further year of research leave that was crucial to my seeing the book to completion. I am grateful to the University of Oxford and to all my colleagues for supporting me during these intense periods of research leave.

The China Environment and Health Initiative provided vital financial support for research in Baocun and Qiancun, but most crucially it also offered the networks instrumental in carrying out fieldwork and data analysis in all three sites. CEHI has worked in close collaboration with Chinese partner institutions through FORHEAD, the Forum on Health, Environment, and Development, which provided an invaluable platform for collaborations in all three case studies, and particularly in Qiancun. Indeed, most of the Chinese colleagues with whom I have worked on pollution and health in rural China were initially contacts made possible by my participation in FORHEAD activities (see the appendix). Jennifer Holdaway, program director of CEHI, deserves a special thank you for skillfully and generously facilitating interdisciplinary collaborations with colleagues in China without which this book would have never been possible. I am grateful for our many stimulating exchanges over the past decade and for her careful and thoughtful reading of my writing over the years. Chen Ajiang, who I met through FORHEAD activities, also deserves a special mention. His diligent in-depth work on pollution in rural China has been a source of inspiration. I am truly grateful to Chen for taking the time to discuss his work on "cancer villages" with me and allowing me to develop my own analysis of case studies originally researched by him and his team, which appear in the second chapter of this book.

During my time in Oxford, I have been fortunate to have the support of many colleagues. Among these are fellow China scholars Karl Gerth,

Elisabeth Hsu, Rana Mitter, Rachel Murphy, Frank Pieke, Vivienne Shue, Tia Thornton, Eileen Walsh, and Xiang Biao, all of whom, in various ways, helped me to refine my thinking. In the school of geography, I am grateful to Andrew Barry, Craig Jeffrey, Linda McDowell, Judy Pallot, and Sarah Whatmore for their stimulating conversations over the years. Peter Wynn Kirby has been a wonderful research companion, providing encouragement and insightful feedback. I benefited greatly from having some excellent graduate students whose projects and thinking intertwined with mine in very productive ways. I am thankful to all members of CHEW (China's Health, Environment, and Welfare research group) for their energy in organizing three conferences so far and creating a forum for debate on these important topics. In particular, I thank Carlo Inverardi Ferri and Loretta Lou who offered incisive feedback on parts of the manuscript.

Beyond Oxford, many colleagues have played instrumental roles at different stages in this project: Deborah Davis, Steve Harrell, Michael Hathaway, Jennifer Holdaway, Sandra Hyde, Arthur Kleinman, Matthew Kohrman, Helen Lambert, Ralph Litzinger, Lu Jixia, Alice Mah, Kevin O'Brien, Bryan Tilt, Benjamin Van Rooij, and Rob Weller all prodded me to develop my ideas further. I am grateful to Fang Jing, Jennifer Holdaway, Lewis Husain, Lu Jixia, and Wang Wuyi for their close reading of chapter 4. I am particularly indebted to Tom Johnson, who patiently, meticulously, and promptly read drafts of various parts of the book. Our ongoing collaborative project on resistance to incineration and rural–urban coalitions has helped to finetune my analysis for this book. Joan Martinez-Alier's enthusiasm and interest in my work, and the opportunity to be involved in his project to map global environmental injustice, widened my geographical horizons and helped me understand China's plight in a broader, comparative context (though much of this will have to be saved for another book).

Different incarnations of parts of this book have benefitted from being presented internationally. In the UK and the rest of Europe, I spoke at the Universities of Bristol, Glasgow, Northumbria, Oslo, Oxford, Plymouth, Sussex, Warwick, and Westminster, as well as at SOAS, the Rachel Carson Center for Environment and Society (Munich), the Max Planck Institute (Halle), the Rockefeller Foundation Bellagio Center, and the Ernst Struengmann Forum (Frankfurt). In China, this work was presented at: FORHEAD conferences in 2009 and 2012 (Beijing), Shantou Medical College, and Hohai University (Nanjing). In the US, it was presented at: Yale and Harvard

Universities, the Association of American Geographers Annual Meeting (Seattle 2011) and the Association of Asian Studies Annual Meeting (San Diego 2013). I am grateful to all who attended these events and raised challenging questions.

For research in Baocun, I am immensely grateful to Zhang Yiyun who capably coordinated, arranged, and took part in fieldwork, and to the dynamic Yunnan-based NGO for which she works, YHDRA (Yunnan Health and Development Research Association), which hosted the project. Thank you to Wu Yunmei, who assisted with her admirable fieldwork skills, to Benjamin Van Rooij and Wang Qiliang, for supporting the project at crucial times, and to all other researchers involved in data collection. Benjamin deserves special thanks for initially introducing me to the site, where he had previously carried out his own in-depth fieldwork, and for providing the contacts that made fieldwork there possible.

For work in Qiancun, I am indebted to Jennifer Holdaway at CEHI, and to Wang Wuyi, Yang Linsheng, and Li Yonghua at the Institute of Geographic Sciences and Natural Resources Research (IGSNRR, Chinese Academy of Sciences). They established a unique collaborative relationship with the county government that was instrumental in arranging fieldwork. In Fenghuang, I am grateful to Teng Zhuren and Peng Zhuren—at the county Centre for Disease Control—who were outstanding in their patience and support for our research, and to the Qiancun village doctor for hosting me during my fieldwork. Lu Jingfang was a dream research assistant, and several other students collected helpful additional material and transcribed interviews. Lu Jixia was an ideal research partner, with whom I thoroughly enjoyed trading fieldwork experiences and preliminary analyses.

For research in Guiyu, I am hugely indebted to Li Liping at Shantou Medical College, who made fieldwork possible by mobilizing contacts she had established in the area through her previous research. Professor Li also introduced me to several students who helped with data collection. In particular, I thank Chen Xuanna and her family for their support. Peter Wynn Kirby and Loretta Lou also provided valuable assistance to my fieldwork during 2012.

My deepest gratitude, of course, is to the research participants across the fieldsites, who devoted so much time and energy to share their experiences with me, even when they felt this may be of little help to them. I hope that my accounts and analysis here do at least some justice to them, even if I cannot give them a voice as such or help to decrease their suffering.

I am grateful to everyone at MIT for their support, particularly to Beth Clevenger and Ginny Crossman for their professional, expeditious, and patient guidance through the editorial process, and to Robert Gottlieb for including the book in his excellent series. Three anonymous reviewers provided extremely useful comments. Ailsa Allen at Oxford offered expert assistance in producing a map.

Last but not least, I owe to my family more than I can put into words. In their own ways, my parents have helped to forge my sense of civic duty, which was a vital ingredient in the inception and completion of this work. My father Roberto worked in a foundry for four decades, witnessing some severe, and yet routine pollution first hand. His keen interest in my project—whether by sharing his experiences or sending me short news articles about environmental injustices in other parts of the world—fueled my sense that this was an important book to write. My staunchly idealistic mother, Liliana, dedicated much of her life to teaching and to the cooperative movement. Through small everyday acts, as well as bolder moves to challenge ignorance, intolerance, and injustice, she has inspired me to take on challenges and persevere when the going gets tough. My sister Elena offered crucial help with finalizing the bibliography and affectionate encouragement throughout. Her dedication to helping vulnerable people through her advocacy work is a strong reminder of what should matter the most in life.

My partner Leon has been a patient, supportive, and inspiring soulmate for almost two decades. He believed I could see this work to the end, even when my own confidence faltered. He was the first to encourage me to travel to China, and indeed it was our trip there together that sealed my lifelong interest in this complex, fascinating country. Thank you Leon for the companionship, the rants, the laughter, the shared dreams. The birth of our son Robin in 2014 has powered me through the final leg of the writing, in an effort to make every hour I spent away from him worthwhile. The arrival of our son Luca just months before the book is published fills me with a renewed motivation to raise questions about injustice, about the uneven burden of pollution, and about the responsibilities governments, businesses, and individuals share in tackling these problems.

# Introduction

In the summer of 2010, I was excited to have the first opportunity to carry out fieldwork in Qiancun, a heavily polluted lead and zinc mining village in Western Hunan province (see chapter 4).[1] For three years I had been interested in how residents of heavily polluted areas in rural China made sense of pollution and its effects, and how they responded to them. The previous year, I had done some similar research in Baocun village, Yunnan province (an area affected by phosphorous mining and processing) with support from the first round of the "China Environment and Health Initiative" (CEHI) grants by the Social Science Research Council (SSRC). In that case, the research was a collaboration between social scientists (including myself) and an NGO (the Yunnan Health and Development Research Association), enlisting help from an epidemiologist and public health specialists. We had collected extensive data on how local residents perceived pollution, and on the rise and fall of local activism (see chapter 3). However, due to the sensitive nature of the project and the tentative nature of our local connections, we had decided against collecting scientific evidence of pollution in the area, such as soil and water samples. In many ways, the accounts we gathered and our own immediate reactions during fieldwork—headaches, skin rashes, and nose bleeds—seemed quite evidence enough of the seriousness of pollution. In addition, research at the main village clinic revealed records of illness that the epidemiologist on our team confidently attributed to phosphorous processing. Villagers, however, did not seem to share the epidemiologist's confidence. The most common attitude toward pollution among locals was a mixture of resentment, resignation—voiced through the ubiquitous expression *mei banfa*, meaning "there is no way" (to end pollution)—and uncertainty over what the effects of pollution actually were. In the absence of data on soil and water contamination, our project was not in the position to prove the extent of pollution or its effects.

Qiancun was different and potentially unique. For several years, medical geographers from China's Academy of Sciences (CAS) had collected blood and hair samples, as well as soil, water, and crop samples, to examine local people's exposure to heavy metals (see chapter 4). This would, I hoped, emancipate me from the question of "but how bad is pollution, really?"— which I was often asked when presenting the Baocun case—by providing some quantitative data on exposure. CAS scholars had established a remarkably close connection with the county's Center for Disease Control, which, at least for some time, granted access to the fieldsite without the complications I had faced in Baocun. I also expected this would mean that the local population would be more confident and more outspoken as a consequence. To better grasp the social, political, economic, and cultural context surrounding Qiancun's mining activities, the China Environment and Health Initiative supported a collaborative, interdisciplinary project under the aegis of FORHEAD (Forum on Health, Environment, and Development). I joined this team as the anthropologist, with additional support from the British Academy. Social scientists on the research team carried out a week-long pilot visit in July 2010, and I stayed behind to conduct more extensive fieldwork with Dr Lu Jingfang, a rural sociologist who capably helped me negotiate the thick local accent (see chapter 4 and the appendix for a fuller description of the project).[2]

On the morning of August 5, 2010, the accountant for Qiancun administrative village, a direct and sometimes intimidating woman in her late thirties, came to the village doctor's house in the subvillage of Guancun, where Jingfang and I lived during fieldwork, and offered to take us to the subvillage of Fengcun for some interviews.[3] Fengcun is a short walk from Guancun and it is downstream from the main lead and zinc mining area, therefore effects on environment and health are most acute there. We had already begun to do some interviews in Fengcun without the help of village officials over the previous days and found the locals as a whole forthcoming, outspoken, and critical of mining. This occasion would prove rather different. Once we arrived, the village accountant asked a woman who was sitting with others by the roadside to promptly return to her house to be interviewed by us. Fengcun's village head, a tall, thin, and seemingly reserved man in his mid-sixties also joined us. Predictably, we had quite some trouble getting her to say anything. The answer to most questions was "I'm not sure" (*bu qingchu*).

We soon moved to the next home. Aunt Lin, a friendly but wary woman in her early fifties invited us in. The house had been built five years before with earnings from a mine her husband had opened, and where, we later found out from others, he had died when he had been crushed by falling rock. The furniture was above average for the village, with lacquered wooden sofas and armchairs, a new flat screen TV and a fridge. Her two sons, both in their late twenties, drove a lorry to and from the mines when business was good. But since 2008 the business had come to a virtual standstill. Our attempts to engage Aunt Lin in conversation beyond superficial answers were met with the same stonewalled standard answers. Under pressure from the village officials to do the interview and not waste their time with pleasantries, we raised some basic questions: "Does mining have any impact?" *I'm not sure.* "Does it affect the water?" *I don't remember.* In the hope of engaging our subject more directly, we tried a leading question I am otherwise keen to avoid: "Does mining affect health?" *I'm not sure. I've never thought about it.* Having been told she suffered from rheumatism and having heard many other locals on the previous days suggest a direct link between that ailment and mining, we tried to make the question more specific: "How do you think you've developed rheumatism?" *I have no idea.*

Frustrated as I felt with the stale situation at hand, I brightened up when a familiar face came into the room. The previous day, Haiwei, a man in his early forties, had approached us as we walked along the road to denounce the pollution of the main well in Fengcun, asking us to expose this in the media or raise it with the central government. But on this occasion he too was reluctant to speak. He emphasized that it is difficult to know if mining has had any effect and that there is no scientific evidence of it. He added that mining is good for the area and that it is the only way to make money. After listening to a morning of uncertainty and subdued silence, Jingfang and I thanked the village accountant and subvillage head for their help in finding interlocutors, and they left us to our own devices.

Once the officials left, both interviewees changed their tune radically. Where Aunt Lin had denied even thinking of any health effects from mining, she now offered several examples of illness and death she believed were caused by pollution from mining. She told us that in the past the village doctor, who died years before, was called upon every day. She also claimed that her rheumatism was due to water pollution, and that so were frequent cases of kidney and gall bladder stones among villagers. Both of them told us that the main well in Fengcun used to be famous in the area for its plentiful fresh water full of tiny shrimp, but that it had been severely polluted and all the shrimp had died. Nevertheless, they argued, all villagers still drink from it. In the half hour we spent by the well, two old men came to collect water and two more came to wash a little further downstream. Interviews with other villagers and participant observation over the course of the following month and during later visits suggested that some of the claims we heard that day were at least partly untrue. For instance, not all villagers drank from this well: many have made efforts to seek other water sources and dug their own wells. Similarly, although these two villagers claimed all the shrimp had died, we could still see some swimming in the well.

The complex nature of these exchanges is telling of the experience of pollution in rural China. Tempting as it might be to regard the exchange in the absence of officials as more valuable or "truer" than responses voiced in their presence, both of them are significant for a social scientific approach. The uncertainty mantra—as I came to call it during fieldwork—is important evidence of how claims about health or environmental damage are politically sensitive, and how such uncertainty is rooted in uneven power relations. Seeming overconfidence and exaggerated claims made in the absence of officials are the reverse of the same coin: attempts by villagers to attract attention and obtain redress while they have largely come to accept that the likelihood of doing so is limited.

This prominent sense of curtailed agency is not the conclusion I expected, much less hoped, to reach in 2007 when I first set out to research how Chinese villagers understand pollution and how they respond to it. One of my key aims was to understand the villagers' potential role in stopping pollution. My research revealed much more complex dynamics. Villagers' reactions may range from violent opposition to industry to acquiescence, with strategic reframing of complaints and requests for compensation. This book then outlines the uneven and nonlinear development of villagers' perceptions and practices related to pollution: how villagers understand pollution and its effects, how they cope with it, how they seek legitimacy for what they consider to be evidence of harm (sometimes seeking help from NGOs and the media), and how their efforts are often frustrated by local governments, by insufficient scientific evidence, as well as by their own sense of powerlessness. It does so by examining closely three severely polluted villages where I carried out fieldwork and several more "cancer villages" (clusters of high-cancer incidence correlated with pollution) drawing on excellent work by Chen Ajiang and his colleagues (Chen et al. 2013).

Pollution is one of the most pressing issues facing contemporary China and among the most prominent causes for unrest. According to a report by China's Academy of Social Sciences, pollution triggered half of the "mass incidents" recorded between 2000 and 2013 (cited in Steinhardt and Wu 2015). The Ministry of Environmental Protection's official recognition of the existence of "cancer villages" in January 2013 (Ministry of Environment 2013a) testifies to the gravity of the problem, not only as an environmental threat, but also as a challenge to social stability, which is a key aim of the Communist Party and one of its core legitimizing principles. The commitment by China's new leadership to building an "ecological civilization" and to wage a "war on pollution" may be signs of change, yet complex problems may not be solved overnight, nor can the overwhelming focus on economic development be easily reversed.

Indeed, this development strategy continues to have deep effects not only on local political economies but also on the ways in which villagers relate to polluting enterprises. Residents of formerly poor, but rapidly developing and industrializing, areas suffer disproportionately from weak pollution regulation. There, local governments face hard tradeoffs between long-term sustainability and short-term needs to provide employment and support public services. In this context, environmental regulations are largely overlooked because polluting firms provide employment and pay taxes (Tilt 2010). This happens largely with the acquiescence of locals who rely on such firms for employment, raising troubling questions over their potential for aiding environmental protection. As a consequence, in areas where villagers' role as whistle-blowers is most important, obstacles to community-based regulation are also most acute: these communities are particularly vulnerable to pollution and least able or inclined to oppose it. Two of the case studies in this book—Baocun and Qiancun—are examples of this predicament.

The conditions under which local communities may oppose pollution remain poorly understood. To be sure, existing studies have highlighted a range of factors that serve as preconditions for citizens' action against pollution and contribute in making community-based regulation successful. These include support from the state and civil society organizations, relatively high incomes and education, independence from local industries, and capacity for organization (cf. Munro 2014). Qualitative case studies, however, show that this list of factors is not exhaustive, nor do they influence citizen action in any straightforward way. Villagers have intricate relationships to pollution, and therefore their attitudes and reactions are complex. To understand these complexities, it is necessary to undertake a more anthropological study of how villagers experience and make sense of pollution, what socioeconomic and political relations exist among communities, local officials, and polluting firms, how patterns of action and inaction develop, and how they relate to shifting definitions of health, environment, development, and a good life. This ethnographic lens offers insights into the complex dynamics of popular contention, environmental movements, and how they relate to local and national political economies.

Several statistics about pollution in China are readily available and much reproduced by the media. But what are the human stories behind them? How do those who live with pollution on a daily basis feel about it? What drives (or obliges) them to stay? How might their experiences, concerns, and sense of entitlement to a healthy environment shift over time? Describing a likely widespread scenario across much of industrialized rural China, this book provides a window onto the staggering human costs of development and the deeply uneven distribution of costs and benefits. Overall, it portrays rural environmentalism and its limitations as prisms through which to study key issues surrounding contemporary Chinese culture and society, such as state responsibility, social justice, ambivalence toward development and modernization, and some of the new fault lines of inequality and social conflict they generate.

## Living with Pollution and the Changing Parameters for a Good Life

Rural and urban dwellers inhabit hugely diverse positions in relation to pollution and they respond to it in disparate ways (see Lora-Wainwright

2013b). A growing number of media reports have described urban middle-class protests against pollution occurring in, among other places, Xiamen, Shanghai, Dalian, Shifang, Qidong, Ningbo, Kunming, and Maoming. With the exception of protests in "cancer villages," rural attitudes and responses to pollution have been less visible. This gives the impression that urban China is witnessing the rise of a widespread environmental movement that remains unmatched in rural China. This belief often goes hand in hand with the assumption that villagers are either ignorant of pollution's effects or that they do not care. Conversely, where rural activism is acknowledged, it is portrayed as a David-versus-Goliath struggle between communities unconditionally opposed to pollution and local governments and industries in cahoots to maintain the status quo, with the latter almost inevitably winning over the former.

This book challenges these assumptions in several ways. First, it shows that, contrary to appearances, villagers' knowledge of pollution and understanding of its effects is often complex and multi-layered. In other words, a relative lack of action is not a simple consequence of lack of knowledge. Neither do they accept pollution in their vicinity only out of ignorance or out of self-interested economic cost–benefit calculus (though economic considerations are of course important, as the case of Guiyu will show in chapter 5). Their resignation to pollution is due to more complex social, cultural, and political reasons, as well as to the intricate relationships they develop with polluters and with the local state over time (Chen et al. 2013; Tilt 2010; Van Rooji 2006; see also Horowitz 2012). In this context, the underlying (and growing) awareness of pollution's harm may escalate into violent protests when particular episodes (acid leaks, explosions, or other severe events) bring it to the fore. At the same time, the longer pollution continues, the more villagers come to regard it as inevitable, and learn to adjust their expectations and demands accordingly. It is to the fluid interplay of activism and resignation that this book turns.

Second, and related, this book puts forward a more emic analysis of how villagers themselves evaluate (in nonrational choice terms) the costs and benefits of the development often coupled with pollution. The result is a co-opted environmental health consciousness: an awareness of pollution's harm on environment and health mixed with disempowerment to oppose it. This is a poignant manifestation of environmental injustice: not only do villagers live with pollution, but they do not feel entitled to demand any better. It highlights the difficult compromises those who live in the shadow of industry have to make. Yet, if we are to understand their experiences more fully, we need to interrogate how their parameters for a good life may have come into being, rather than emphasize injustice from a normative perspective. Indeed, for those who live with contamination, toxicity is part of their natural environment, what I refer to as "toxic natures." If pollution is part and parcel of nature, its bodily effects also come to be regarded as normal. Over time, local populations adjust their parameters for a good life to include pollution. With this in mind, their experiences of pollution and their responses need to be understood within the broader context of the many other challenges they face. Against purely economistic analyses, this book shows that forms of engagement with pollution are not only economic decisions but also deeply social and moral ones.

Third, the book looks beyond high-profile cases of successful resistance and focuses on the much more common scenarios in which pollution victims suffer in silence, are unsuccessful at ending pollution, are co-opted into seeing it as inevitable, or draw economic benefit from polluting activities. Recent scholarship has rightly highlighted the importance of the media, litigation, and environmental NGOs as tools in citizens' environmental struggles (see chapter 1). Much of social science research on environment and health in China to date focuses on localities and cases where environmental threats have fairly clear health effects and have led to contention (Holdaway 2013). By contrast, although most environmental suffering takes place far from the purview of journalists, courts, and NGOs, the daily grind of "living with pollution" has received scant attention. As a consequence, with the exception of "cancer villages" (which have succeeded in attracting attention), rural environmentalism is more likely to remain invisible. This book attends to these less visible forms of environmental suffering. This requires a rethinking of existing environmental justice models, to better grasp the origins of resignation and scenarios in which the line between victims and beneficiaries is blurred. In doing so, the book develops a more nuanced perspective on citizens' agency and revisits concepts drawn from collective contention and comparative environmental justice. The concept of resigned activism may be useful to understand the nuances of their experiences and reject romanticized notions of peasant resistance.

Fourth, focusing on single strategies—whether they be resorting to journalists, scientists, or other "experts," NGO campaigners, petitions, litigation, or violent protests—only produces a partial view of how communities experience pollution and deal with it. This book by contrast zooms in on three particular communities and examines their attitudes and responses in a longer time frame. At times, villagers may become staunch defenders of polluting businesses and, conversely, at other times, they may display great resilience and creativity in their opposition to pollution. A detailed account of each strategy embraced in the three sites is beyond the scope of this book—indeed, this would require an entire monograph devoted to each site. Instead, the book highlights modes of "living with pollution" that would be less visible if the emphasis remained on collective action. It accounts for how concerns are shaped, how entitlements change, and how those who live with pollution learn to adjust their expectations and their demands. It maps out an uneven terrain in which citizens are concerned with environmental health threats, are diversely positioned to overcome them, and embark upon varied pathways of action to protect themselves individually, as a family, or as a community. In turn, villagers emerge not as stable subjects, but as involved in ongoing processes of negotiation with their families, neighbors, the polluting firms, various levels of the state, and a range of outsiders.

## Resigned Activism

The seeming oxymoron at the very center of this book—resigned activism—is perhaps best elucidated by my experience in Baocun village (see chapter 3). Baocun was originally selected as a fieldsite based on previous excellent in-depth research in the field of political and legal studies that suggested there had been limited instances of environmental activism in the area (Van Rooij 2006). The premise of our fieldwork there was to understand why a community living with severe contamination did not oppose it. The research team and I were predictably surprised, therefore, when we discovered that Baocun had actually experienced decades of activism (albeit localized in nature), starting soon after the local industry began to operate. Why was there such a discrepancy in our findings? Different methodologies and interview techniques go some of the way to answering this question. Researchers' own biases about how activism is defined and what may be included of course influence the way we raise questions about it and the answers we receive. The open line of questioning we adopted during field-work allowed research participants to interpret questions such as "what can be done" about pollution as they chose and elicited descriptions of past protests, petitions, and blockades that they may not have volunteered oth-erwise. Specific questions about whether locals engaged with the political process or resorted to legal pathways in response to pollution may not have triggered these accounts because (as we came to learn) locals discounted these efforts as futile. By contrast, a more open discussion about pollution in the context of everyday life enabled research participants' dissatisfaction and sense of frustration to surface more easily, even where they may have felt that their responses had limited outcomes.

Most crucially, however, this discrepancy is derived from a conceptual blurring between resignation and activism apparent in all the fieldsites. Indeed, the fact that Baocun could seem both acquiescent *and* activist pre-sents an important lesson: surely the boundary between these two attitudes and forms of engagement is not as clear as we may have assumed. An anthropological sensibility to the primacy of locals' own categories and experiences allowed this blurring to emerge clearly during fieldwork and subsequent analysis. In turn, the co-presence of resignation and activism demands that we situate the whole spectrum of attitudes and reactions to pollution vis-à-vis complex, shifting, and uneven social, cultural, political, and economic contexts. Why does such blurring take place and with what effects?

The expression *mei banfa*, meaning literally "there is no way," was the most common reply I received when, in the context of discussions of pol-lution with villagers in all three sites, I asked the open question "what can be done?" (*zenme ban?*). It was so prominent that I considered adopting it as the first part of the book's title. Ultimately, I decided against it because a simplistic reading of this statement would result in an incomplete charac-terization of the more layered entanglement of perceptions and practices at play. Upon an initial and superficial analysis, the frequent declaration *mei banfa* may seem to portray a situation in which subjects are determined by their contexts, where social reproduction and structural constraints leave no room for agency. The implications of these statements, however, are more complex than they may at first seem. As uncompromisingly defeatist as *mei banfa* sounds, it should not be taken at face value. *Mei banfa* does not literally mean that people have done nothing and intend to do nothing to engage with pollution. It does not mean that their actions have had no effect, or that they necessarily *believe* that their actions have no effects at all. Rather, it is a way to convey their own feelings of powerlessness: to emancipate themselves from the expectation (from themselves, their family, the community, the anthropologist) that they may be personally responsible for effectively curbing pollution. It is a means through which they comfort themselves about the limits to their agency.

In his 2010 book, Steve Lerner examined a dozen cases of what, following environmental activists, he calls "sacrifice zones": locations across the United States that are disproportionally affected by pollution. He explained that residents of sacrifice zones endured severe pollution until their "rude awakening" (Lerner 2010, 8) prompted by events such as "the discovery of a cluster of pollution-induced illnesses or the release of a report or newspaper article revealing the extent of the contamination" (9). This in turn galvanized residents to organize and resulted in grassroots environmental justice struggles. Phil Brown (2007) described similar processes underlying the development of "embodied social movements" and "popular epidemiology": individual awareness of pollution is followed by the social discovery of a disease cluster and the politicization of the local community. Similar dynamics are documented in the Mississippi chemical corridor (Louisiana) that has come to be known as "cancer alley" (Allen 2003; Lerner 2005; Ottinger 2013) and in other locations across the US (Brown, Morello-Frosch, and Zavestoski 2012; Kroll-Smith, Brown, and Gunter 2000; Little 2014), and beyond (Auyero and Swistun 2009; Das 2000; Fortun 2001; Mah 2012; Kirby 2011; Petryna 2002).

By contrast, this book shows that there is no inevitable linear development leading from the discovery of pollution's detrimental effects on the environment and health to the formation of collective identity, the politicization of the local community, and the emergence of citizen-expert alliances. Routine pollution and acute events (such as the explosions in Baocun) alarm local residents and may sometimes prompt localized political acts such as small blockades, protests, and petitions. However, such acts rarely escalate to higher levels (with the exception of some petitions in Qiancun, as we shall see in chapter 4), infrequently involve obtaining the support of scientific experts, or succeed in attracting substantial redress. Studies of environmental injustice understandably tend to focus on the mechanisms through which activism develops and to examine its effects. Less attention is given to fatalism, resignation, and to the processes through which pollution becomes rooted in local communities. Alternatively, where the absence of sustained collective action receives attention, it is often explained as a direct consequence of the locality's economic dependence on polluting activities. Against these trends, this book asks: What about communities that instead of coming together against pollution are torn apart by it? What about a wider range of activism that includes its less visible forms? What about resignation? How is resignation shaped by dynamics beyond simple economic dependence? How does uncertainty of the kinds described in the opening vignette come into being and with what effects?

The term "resigned activism" serves as a conceptual tool to attend to subtle shifts in parameters and expectations, and to the diverse forms of environmental engagement they support. It encapsulates a spectrum of perceptions and practices comprising acts that may fit the conventional label of collective environmental contention, such as protesting at the factory gates and filing petitions. But it also includes less confrontational and more individualized or family-oriented strategies aimed at minimizing pollution in one's immediate surroundings: closing the windows at night to limit exposure to fumes; wearing masks; buying bottled water, at least for children; sending children and pregnant women to live elsewhere; quitting the most harmful jobs; or discursively defending one's own work as less harmful than that of others. Of course, most of these actions do not amount to collective resistance. Bryan Tilt (2013) described similar patterns as part and parcel of a wider trend toward "individualisation" in Chinese society (Yan 2009), ways of countering pollution on an individual or family basis rather than as a community (see also Lora-Wainwright 2013d). Conventional approaches to environmental activism would neglect these practices as inconsequential by virtue of their limited political momentum and absence of collective spirit. However, the seeming absence of collective action against pollution should not be a reason to ignore subtler, less visible forms of environmental engagement. These practices may not present the characteristics of a full-fledged environmental movement, but their effects are no less significant, even if those effects are more on the individuals themselves than on the context they may have originally wished to change.

In this guise, activism comprises the small steps individuals and families take in order to minimize the physical, psychological, and social effects of pollution. It includes individual efforts to alter one's most immediate circumstances (the air inside one's home, for instance) and to protect one's family, but also shifts in attitudes and expectations to accommodate and normalize pollution. The power and effects of these processes of attunement may only be grasped through an examination of the wider challenges people face and how they intersect with the gradual and deep embedding of pollution. This requires a holistic approach to health, environment, and development as they intersect in locals' lives. Rather than assuming a normative definition of healthy environments and healthy bodies, researching resigned activism in severely polluted localities demands that we take seriously locals' own diverse languages of valuation (see Guha and Martinez-Alier 1997), and what has come to be considered as a healthy—or at least a bearable—environment in the context of toxic natures. These adjusted expectations, rather than economic dependency alone, powerfully affect the forms that activism may take.

This suggestion has broader conceptual implications for the study of environmentalism. It demands that we attend to environmental concerns and environmental activism that may be present in unexpected and less visible forms. Indeed, it requires a redefinition of activism that would allow a study of these less obvious forms of engagement. This approach entails combining attention to resilient forms of activism and the often slow process through which environmental plight may become embedded and normalized, or, to borrow Gramsci's (1971) terminology, "hegemonic." However, as this book shows, the hegemonic nature of pollution is not static. Following Raymond Williams' (1977) influential critique of Gramsci's concept: "A lived hegemony is always a process ... It is a realised complex of experiences, relationships and activities, with specific and changing pressures and limits ... It has continually to be renewed, recreated, defended, and modified. It is also continually resisted, limited, altered, challenged by pressures not at all its own" (112). Similarly, resigned activism and living with pollution encompass complex, multiple, diachronic, and contested processes through which pollution is both challenged and naturalized, as locals' parameters and their expectations shift.

The process of altering one's demands in response to circumstances was perhaps most famously studied in social science circles by French sociologist, anthropologist, and social theorist Pierre Bourdieu. Throughout much of Bourdieu's extensive corpus of scholarship, he elaborated conceptual tools such as the *habitus*, taste, and the "feel for the game" (1977, 1984, 1990), which in various ways make sense of the processes of social reproduction, of the ways in which individuals are complicit in entrenching their own social position by taking their entitlements (or more pertinently, their lack of entitlement) for granted and adjusting their expectations to the *status quo*. Bourdieu refers to these dynamics as "symbolic violence." The term describes subtle and implicit processes of naturalization and legitimizing of unequal power structures (1991, 51). Through symbolic violence, subordinate social groups are rendered doubly powerless: they are in a structurally disadvantaged position and they come to accept their inferiority as normal (1990, 118).

Bourdieu has been often criticized (unfairly, at least in part) for presuming too much stability and precluding the possibility of agency.[4] While his interests lie predominantly with processes of social reproduction rather than social change, his attention to subjects' practical logic—similar to players who learn to anticipate where a game might go and to follow its rules—also highlights flexible forms of agency. The physical—if slow—violence of living with pollution is combined with symbolic violence to the extent that people have learned to take it largely for granted and to adapt their expectations to seemingly inevitable circumstances. But residents of highly polluted sites in China's countryside also attempt to craft a better future for themselves and their families, however circumscribed they may be. Their agency may be tactical rather than strategic—typical of those who cannot transcend their conditions (De Certeau 1984, 37)—but it nevertheless deserves attention.

Processes of attunement to pollution are not only social, cultural, political, and economic, but they are also powerfully and inextricably embodied. Indeed, engagements with pollution affect not only expectations about what counts as a healthy body, but also what a healthy body *is* in the context of toxic natures. Margaret Lock (1993) elaborated the influential concept of "local biologies" to make sense of cultural and biological differences in experiences of menopause in Japan and North America. She argued that the diversity in symptoms reported among women in these two contexts were not only to do with culturally specific understandings of menopause, but also with biological, embodied differences, potentially rooted in different diets. The concept of local biologies suggests that culture and biology are coproduced; they both shape experience and this in turn shapes discourses about the body.

Where Lock noted the development of local biologies that are spatially discrete, this book portrays local biologies that develop diachronically in the three research sites. Locals reported that symptoms that affected them at first (for instance, nose infections in Baocun and Guiyu) no longer troubled them after some time. Migrants were presented as the clearest evidence that one can "get used to pollution," as their initially acute symptoms abated with time. The alleged decrease in symptoms may be to do with bodies' attunement to their surrounding toxic natures. Perhaps the loss of some respiratory function due to chronic and acute exposures resulted in less prominent symptoms. Exposure to pollution rendered local residents seemingly immune to its effects. But such physical immunity is also inextricably psychological and social: chronic and common symptoms ceased to seem significant. Local biologies in these severely polluted sites consist of bodies *and* minds attuned not to notice pollution's effects, not to dwell on them because they are widespread and seemingly inevitable. These shifts in perception, experience, and expectations are part and parcel of the forms of resigned activism described in this book.

## Limitations and Reflections

Carrying out research in China as a foreigner has its complications. On the most basic level, needing a visa and other approvals by local governments overseeing the research location means networks need to be in place and sometimes lengthy negotiations need to be undertaken in order to secure access. Of course, Chinese scholars also face many of these difficulties. Being physically more remote from fieldsites than Chinese collaborators and colleagues adds a further burden when it comes to coordinating research schedules and sharing analytical perspectives—something that is much more easily done in person or at least with the benefit of geographical proximity. When research concerns highly sensitive topics such as rural pollution, health, and activism, these complications increase significantly. Access to fieldsites has to be negotiated carefully and patiently, and may in fact never materialize, or cease to be feasible at certain times, as was the case in Qiancun.

While Chinese researchers face similar obstacles in researching sensitive topics, the consequences differ markedly. At worst, a foreign researcher may be deported and denied access to the country in the future. Chinese nationals may face more severe repercussions. Therefore, participating as a foreigner in projects that involved Chinese colleagues and collaborators in the respective local governments required added care to minimize the chances of my involvement jeopardizing their positions. This meant that the relative freedom with which I undertook fieldwork previously, even on sensitive topics such as forced relocation, was considerably curtailed.

Leading or participating in research projects populated largely by Chinese colleagues has put me in a position of relative privilege when it comes to analyzing the findings and reflecting on these projects in English-medium publications. I am keenly aware of this imbalance. To address it, I have previously published in conjunction with some of my closest collaborators (see for instance Lora-Wainwright, Zhang, Wu, and Van Rooij 2012; Lu and Lora-Wainwright 2014) and have sought to consult colleagues as I revised this book's manuscript. Nevertheless, I make no claim to offer an objective account encompassing the views of all colleagues who took part in these projects. Interpretations remain, of course, my own, and I am fully responsible for any omissions or partial representations.

Last but certainly not least, I am painfully aware of the imbalance between those I describe in this book, who by and large continue to suffer from pollution, and the academic standing I may derive from writing about them. For those who endure severe pollution, reflecting on its effects can be very excruciating and disheartening. For me, it was often heartbreaking to listen to their stories, their anger, their disillusionment, and their hopes for a better life without being able to offer comfort or help, whatever that might mean. I open this introduction and the following three chapters based on my case studies with personal reflections and notes from the field, in the hope of taking the reader closer to the experience of "being there," which is so central to anthropology. But I made a conscious decision not to take a more prominently self-reflexive stance throughout the book. This is not to deny that I am the architect of the arguments I put forward, or to pretend I have produced an objective account of what I witnessed. Rather, I do so to avoid my own experience taking center stage and detracting from the rightful subject matter: the experience, suffering, and practices of those whom I describe.

## Chapter Summaries

Chapter 1 situates the book within a broad range of relevant literature. In the first part, it engages with social science of environmental health and environmental movements, particularly about the study of social movements, varieties of environmentalism, complexities in environmental health, and the role of citizens in contesting them. On this basis, it proposes a diachronic focus beyond single strategies and attention to the processes through which environmental concerns are formed and their complex interplay with local contexts. It calls for a more emic approach to how local communities value environment and health within the broader context of their everyday lives, and how and why these values may change over time. It presents "resigned activism" as an analytical tool for bridging analyses of activism and resignation, and for showing how they merge across a wide range of villagers' attitudes and everyday practices. In the second part, it outlines China's major environmental challenges and rising forms of environmentalism, including a broad set of actors. It suggests the need to shift the focus away from formalized and organized environmental activism—resulting in numerous studies of NGOs—and to examine more routine and often invisible forms of environmentalism emerging in rural areas, as well as the reasons why such concerns may be downplayed or silenced.

Chapter 2 examines the emergence of China's "cancer villages"—village-sized clusters of high cancer incidence—and their significance. It overviews how media accounts discursively shaped their social, political, and epistemological nature. It develops a typology of "cancer villages" based on a close analysis of a selected number of cases examined in recent qualitative research by the leading sociologist of rural pollution in China, Chen Ajiang, and his team (Chen et al. 2013). These relatively high-profile, politically active cases provide a useful background against which to compare the less visibly active case studies examined in later chapters. They illustrate a broader range of activist practices, but they also show that such strategies are often ineffective. Ultimately, these examples suggest that the term "cancer villages" is not an epidemiologically uncontested label, but rather a cultural, social, economic, and political phenomenon. Further, they prove that scientific evidence is not the most important element in gaining redress. Rather, it is the socioeconomic contexts, the persistence of the local population's complaints and their ability to threaten social stability, which largely determines the ways in which polluting firms and the local government may respond. This point is further supported by the book's three case studies, in which scientific evidence plays a relatively minor role in villagers' reckonings about environmental health effects and in their demands for redress.

Chapter 3 begins to flesh out the contours of resigned activism through the case of Baocun village, a major site for phosphorous mining and fertilizer production. It shows how industrialization deeply diversified the local population, ranging from poor migrants to wealthy business owners, and bore unequal effects on them: while some are better positioned to take advantage of opportunities, others suffer a precarious existence affected by socioeconomic marginality and the slow violence of unaddressed environmental hazards upon their health. It focuses most closely on unfolding processes of resignation among the migrant population—who stand to suffer the most from pollution—and poor locals. The chapter illustrates how financial dependence on polluting activities is a form of "disaffective labor" (cf. Hardt 1999) that pushes migrant workers and poor locals to take pollution and their precarious position for granted. They regard toxicity as a part of the natural environment and environmental afflictions on the body as "normal." In this context, the value of life and parameters to define it are slowly but firmly altered.

Chapter 4 describes the evolution of lead and zinc mining in Qiancun village and its effects. Mining contributed to the entrenchment of socioeconomic stratification, as well as to the shifting and uneven effects upon environmental health. It affected livelihood pathways available to the local population, their perceptions of the benefits and effects of mining, and the ways in which they valued the environment. This chapter takes a closer look at how dynamics of resigned activism overlapped with marginalization and accusations of "madness" waged against one of the foremost figures in local activism. It explores the intersections between one issue that particularly troubled locals—provision of safe drinking water—and local politics and the lack of trust in local officials. Locals' efforts to secure safe drinking water also threw into sharp relief their inability to coordinate effectively, which reinforced feelings of resignation. Finally, the chapter elucidates some of the dynamics animating the interdisciplinary project of which my fieldwork was part and some of the elements that shaped what experimental intervention pathways were explored and embraced at an early stage in this ongoing project.

Chapter 5 examines a third case study that differs in important ways from the first two studies. Unlike Baocun and Qiancun, Guiyu town is a well-known, indeed notorious, environmental health hotspot. Pollution is caused by a vast and complex cottage industry processing electronic waste. Chapter 5 explores how such "e-waste work" became closely embedded within the local community, family and social relations, as domestic and work spaces were inextricably blurred. It disaggregates the black box of "e-waste work" to show how it evolved over time, the great diversity that composes the sector, how the government attempted to regulate particular activities within it, and why their efforts were not fully effective. It shows that, as in Baocun and Qiancun, the economic benefits and environmental costs of these activities are unevenly distributed. By describing a range of diverse e-waste workers engaged in a spectrum of more or less polluting work, the chapter illustrates how locals fashion counter-discourses of relative harm to excuse their practices and avoid blame. In these circumstances, as in Baocun, toxicity is naturalized, and parameters of health are adjusted to normalize and accept widespread pollution-induced ailments.

The conclusion draws comparisons across the three sites and it highlights common dynamics and processes, such as the normalization of pollution, and the molding of new parameters of health and new expectations for a "good life." It closes by returning to the main themes of the book and to their implications for the social science study of environmentalism and of contemporary China. It reflects briefly on the wider global responsibility for the forms of pollution and suffering described, on the importance of looking beyond conventional forms of activism, and of taking local contexts seriously. It puts forth some suggestions for how academics might contribute to empowering communities affected by pollution.

# 1 Situating the Study of Rural China's Environmental Health Activism

"Pollution affects health. So what? Nobody asked you to be born in this village. Living here has benefits and drawbacks. The drawback is pollution. ... I live as long as fate allows (*huo jitian suan jitian*). ... So if I have to breathe this poison I don't care, that's how we live."
—Fifty-four-year-old woman with a low income, Baocun, May 19, 2009.

## Environmentalism, Health, and Activism

### Social Movements, Resistance, and Acquiescence

The study of social movements and resistance provides a useful broad framework for the study of environmental health activism (below I expand the excellent overview by Klawiter 2008). The field developed significantly over the past half century. The theorization behind early social movements was rooted in positivism and neoclassical economics, and therefore treated actors as rational self-interested individuals who participate when benefits outweigh the costs. This intersected with resource mobilization theory (as exemplified by Tarrow 1983), which emphasized the importance of resources to the rise of activism, but did not examine in detail how and why movements develop. Political process theory expanded this approach by stressing the importance of political opportunities structures to mobilization (McAdam, McCarthy, and Zald 1996). While these approaches focused largely on material and structural elements, later scholarship highlighted that efforts to craft a shared understanding of a problem, that is, how grievances and responses were socially constructed, or "framed," played a crucial role in mobilization (Snow and Benford 1992). Perceptions and ideas, in other words, are important drivers (or inhibitors) of activism (Gould, Schnaiberg, and Weinberg 1996; Jasper 1997). Combining insights from this literature, this book will explore the importance of material and structural elements as they intersect with knowledge formation and framing in order to understand approaches to pollution and forms of action and inaction as they develop over time.

Early social movements theories have been variously criticized for their overwhelming focus on the state as the target of social movements, the scant attention to conflicts within movements, and the lack of a nuanced analysis of the relationship between suffering, subjectivity, and social movements (Klawiter 2008, 13). The turn toward the study of New Social Movements was a reaction to this bias. Partly claiming to reflect the rise of new movements (for instance, those to protect human rights, to advance civil rights, and to promote gender equality), and partly offering a new approach to examining movements as a whole, New Social Movements scholars placed the focus on identity and on the body as a site of contention (Mellucci 1989). Unlike working-class movements demanding redistribution of resources, these movements engaged in the "politics of recognition" (Fraser 1996). The suggestion of a categorical shift from politics of redistribution to politics of recognition however is one-sided. Klawiter's study of advocacy surrounding breast cancer in the US shows that movements may be at once about redistribution *and* recognition; they may target the state alongside other institutions. Similarly, as Phil Brown (2007) shows, those who join "embodied health movements" around asthma, cancer, and Gulf War Syndrome seek recognition, but they also complain about the unfair distribution of hazards. This book takes these critiques on board and illustrates the ways in which these concerns with redistribution and recognition may clash and collaborate. It shows how the dynamics behind the rise and fall of environmentalism have many more counterparts than the state or even the polluting companies. Understanding these dynamics requires a study of community cohesion and tensions, conflicts of interests, and conflicts over the uneven distribution and spread of benefits and harm.

One innovation in the New Social Movement scholarship that is crucial to this book is the focus beyond single strategies, such as protests, to a much wider repertoire of action. This may include protests, sit-ins, court cases, marches, more diffuse communication processes across multiple locales, use of mass media and new technologies, and engagement with scientists, NGOs, and various government departments (Leach, Scoones, and Stirling 2010, 147). These different spaces and modes of mobilization affect the ways in which problems are understood and what is seen to count as evidence of a problem at all. Likewise, growing scholarship on coalitions moves beyond a focus on single tactics and localities to examine "the production of a new connection between previously unconnected sites" (Tilly and Tarrow 2006, 31), contributing to the "scaling-up" of localized concerns into environmental networks and broader social movements (McAdam and Boudet 2012; Saunders 2013; see also Gottlieb 2002). The term "mobilizing citizens" portrays citizens as knowledgeable actors engaged in a dynamic, networked politics, involving "shifting and temporary forms of social solidarity and identification" and local, national, and global networks (Leach, Scoones, and Stirling 2010, 141; see also Leach and Scoones 2007).

A fuller understanding of social movements can only be achieved by also examining what is commonly regarded to be their diametrical opposite— acquiescence—and the conditions under which it is sustained. Italian Marxist theorist and prominent political figure Antonio Gramsci (1971) proposed the concept of hegemony to refer to instances where domination takes place not by force but by consent. Attending to similar dynamics, John Gaventa opened his widely acclaimed study of quiescence and rebellion in an Appalachian valley with the question: "Why, in a social relationship involving the domination of a non-elite by an elite, does challenge to that domination not occur?" (Gaventa 1980, 3).[1] Quoting Steven Lukes (1974, 23), Gaventa explained the workings of power as follows: "A may exercise power over B by getting him to do what he does not want to do, but *he also exercises power over him by influencing, shaping, or determining his very wants*" (Gaventa 1980, 12, his emphasis). Eventually, Gaventa argued, B begins to withdraw unconsciously, due to his or her "sense of its own powerlessness" (17), coupled with a history of withdrawal. The sense of powerlessness is historically formed, through past experiences of defeat. As a consequence, B may not act because he or she has learned to view "the order as immutable or through lacking conceptions of possible alternatives" (20).[2]

Exploring some of the vast sociopolitical space between activism and resignation, James Scott (1985) famously highlighted less visible, everyday "small acts of resistance" such as irony, satire, foot-dragging, and sabotage, which he termed "weapons of the weak." Against Gramsci, he argued that such everyday acts show that subaltern subjects have not fully consented to domination. Like Scott, I am interested in the space between domination and resistance and in challenging their dichotomy. However, instead of only focusing on resistance, I examine various forms of activism alongside resignation and the feelings of powerlessness described by Gaventa. Most crucially, I argue that more attention is required to the processes through which activism is shaped (both in its aims and its forms), as well as to how expectations shift to accommodate resignation. Accounts of the three core case studies reveal that locals' concerns are not always unified and that mobilization efforts are uneven across the socioeconomic spectrum and through time.

## Varieties of Environmentalism

Environmentalism is globalized in two senses: vertically because of concerns with global justice, and horizontally because it emerges in different places (Walker 2012). It involves broader, sometimes globally oriented "not-in-anyone's-back-yard" (NIABY) coalitions concerned with the transfer of hazards to poorer regions or countries in the global South (Bullard 2000; Carmin and Agyeman 2011; Clapp 2001; Pellow 2007). Likewise, those who study environmental justice movements have advocated an analytical framework that sits between militant particularism and theoretical universalism (Schlosberg 2007) and examines the multiple "spaces of environmental justice" (Holifield, Porter, and Walker 2010). This book contributes to these debates by showcasing the development of a silenced, largely invisible environmentalism in rural China. Two of the cases—Baocun and Qiancun—may seem to fit the label of NIMBY (not-in-my-back-yard) movements, but I show that their connotations, structural contexts, strategies adopted, and outcomes are rather different than US-based studies might suggest (e.g., Lerner 2010), raising questions about the applicability of theoretical models based on cases from the global North to the Chinese context. When judged according to normative standards of what may count as success for an environmental movement—the ability to decrease or cease pollution—these practices may seem largely ineffective, but I show that they nevertheless exerted vital effects upon locals' lives.

The idea of "environmentalism of the poor" (Guha 2000, Guha and Martinez-Alier 1997, Martinez-Alier 2003) was put forward in part to tackle differences between environmentalism in the global South and the global North. It suggested that while Northern environmentalism is largely premised on post-materialism (i.e., concerns with the environment only emerge when material needs have been met), Southern environmentalism is premised on regarding the environment as a source of livelihood. This form of environmentalism focuses on social justice, claims to recognition and participation, and efforts to defend indigenous land rights and preserve their livelihoods against mining, dams, land grabs, oil, and gas exploitation (Bebbington, Hinojosa, Bebbington, Burneo, and Warnaars 2008; Bebbington, Bebbington, Bury, Lingan, Muñoz, and Scurrah 2008; Bridge 2004). While it is important to avoid using environmentalism in the global North as a blueprint for environmentalism elsewhere, this portrayal is one-sided for three main reasons. First, Northern environmentalism—to the extent that it can be regarded as a coherent movement—is not always and only driven by affluence or self-indulgence and may also involve opposition to environmentally destructive facilities (Horowitz 2012). Second, environmentalist ideologies and practices do not only travel from the global North to the global South, but rather "environmental winds" move much more fluidly in several directions (Hathaway 2013). Similarly, environmental concepts flow between academia and activist circles and across regions through multiple networks and learning processes (Martinez-Alier 2014; Martinez-Alier et al. 2016). Third, and most relevant for this book, the implication that the poor always resist environmental degradation and protect nature is a romanticized portrayal that ignores that local communities have mixed views of development, and they may be victims of pollution as much as they are complicit in it or even perpetrating it themselves (Kirsch 2007). Celebrating "ecosystem people" as alternative authentic voices may result in ignoring diversity and darker elements of their discourses (Williams and Mawdsley 2006, 668).

This criticism may be leveled at the concept of environmental justice as a whole: to the extent that it relies on self-definitions of victimization, environmental justice may ignore those who suffer from pollution but do not necessarily self-define as victims. This point may be extended to any uncritical study of grassroots environmentalism. The term itself implies a populist movement among the masses against 'big powers' like industry or the state. Yet dynamics are more complex: investors and workers may be among those fighting pollution, movements are full of tensions and conflicts, and they may be better characterized as sporadic and short-term rather than as sustained resistance (Gould, Schnaiberg, and Weinberg 1996). Based on their case studies, Gould, Schnaiberg, and Weinberg argue that opposition to pollution focuses on health concerns instead of development only when there are other options for development, otherwise residents are likely to accept the facility (187–188). This is an important point, but, as this book shows, concerns with health and the environment may continue to coexist alongside the desire for development. Even when residents no longer demand that industry should protect environment and health, their concerns persist and may take the form of more individualized responses to protect themselves and their families from pollution. More crucially, these dimensions of life are not so easily separated. In light of such insights, this book avoids drawing general conclusions as to whether the poor or the wealthy care more for the environment, but rather focuses on how different communities and their members regard the environment and their health, how they define what is worth protecting, what they regard as the problem, and what solutions they envision and embrace. Similarly, the success of environmental action should be assessed according to the aims set by participants rather than against a predefined standard. Their ideas of what may count as success vary through time, depending on the outcomes of previous actions (Lora-Wainwright et al. 2012).

Indeed, one vital contribution of the environmentalism of the poor approach is to highlight that there are different languages of valuation (Guha and Martinez-Alier 1997). Environmental movements sometimes resort to a language of economic valuation (for instance, demanding compensation for externalities), but they may also use the language of sacredness and of territorial and human rights. It is equally vital to examine how justice and fairness are defined. As Amartya Sen put it, "we can have a strong sense of injustice on many different grounds, and yet not agree on one particular ground being *the* dominant reason for the diagnosis of injustice" (Sen 2009, 2). Likewise, definitions of justice vary in different contexts, and depending on whether they are formulated by communities, campaigners, legal systems, or compensation mechanisms (Schlosberg 2007). This demands taking a step back and questioning given definitions, and examining instead how justice, fairness, and moral behavior are defined by villagers who live with pollution. This also enables a more nuanced approach to villagers, not simply as victims, but as subjects who inhabit complex positions vis-à-vis pollution.

Given these complexities, a detailed and diachronic study of the contexts of environmental concerns and environmental movements is crucial.

The broad field of political ecology highlights how unequal power relations produce particular discourses on the environment and subjects endowed with the power and responsibility to safeguard what is deemed worthy of protection.[3] Arun Agrawal's study (2005) of how Kumaon villagers (North India) became party to forest conservation is a classic example of these processes. It outlines the historical and contingent emergence of "environmental subjects" who care for the environment and see themselves as the guards of local forests. In a similar way, this book traces the emergence of different types of environmental subjects across the three sites. These environmental subjects do not necessarily care for the environment in a conventional sense. At particular points in time, they may demand a clean environment, but at others they participate in reconfiguring the value of the environment as a source of profit and a resource to be exploited rather than a communal good to be protected.

To understand these processes, I do not postulate actors with already fully formed opinions, but rather build on Agrawal's work on the formation of environmental subjects to ask such questions as the following (2005, 164): "When and for what reason do socially situated actors come to care for, act, and think of their actions in relation to something they define as the environment?" Also based on Agrawal's work, we may ask how and why, conversely, do people come to view pollution as inevitable? These complexities demand a study of how particular environmental subjects emerged in response to the local context. How are their interests defined? How do they develop over time? What elements influence attitudes to the environment and how do they intersect with attitudes to development? In turn, this approach enables a more nuanced approach to the relationship among villagers, polluters, and the local state. It allows us "to dissect simplistic, essentialized interpretations of environmental struggles, unpacking concepts such as the 'community' or the 'social movement' by examining the ways in which micro-level political, economic, and social interactions influence grassroots visions of industrial development and actions to address it" (Horowitz 2012, 23).

## Environmental Health Complexities

Environmental health is a contested field given scientific uncertainty in both toxicology and epidemiology. Establishing causal links between exposure and health effects (which is often necessary for those who regard themselves as victims of pollution and want to demand redress) is complex for several reasons: there are no scientific studies for large numbers of chemicals, exposure may take place over a long-term period (see Nixon 2011), risk may be posed by a multiplicity of factors rather than a single substance, incidence of chemical exposure is difficult to pinpoint, symptoms may not fit a typical pattern, and experts disagree about the import and even the existence of widespread, low-level exposures (Brown 2007; Brown, Kroll-Smith, and Gunter 2000; Brown, Morello-Frosch, and Zavestoski 2012; Murphy 2006; Steingraber 2010; Tesh 2000). As a consequence of such complex causality and disputed standards of evidence, "virtually all diseases and conditions that can be attributed to environmental causes are highly contested and the source of considerable confusion, anger, and resentment" (Kroll-Smith, Brown, and Gunter 2000, 9).

While environmental health is haunted by radical uncertainty, such uncertainty is often maintained given the political and economic interests involved.[4] Uncertainty in turn reinforces the sociopolitical assemblages that kept it in place. Indeed, while in many cases the copresence of several factors that might precipitate cancer in addition to pollution complicates matters, in others the link between pollution and illness may be relatively clear (Holdaway and Wang 2013). Yet complex overlapping circumstances mean that local populations are unable to secure redress. Several authors, particularly in science and technology studies, have pointed out that industrial and political interests shape science and standards of evidence (Murphy 2006; Waldman 2009, 2011). Such processes are closely tied to what Proctor and Schiebinger (2008) have termed "agnotology," or the production of ignorance, and "undone science" (Frickel et al. 2010). The "toxic uncertainty" produced by these strategies is rooted in uneven power relations and is a form of domination (Auyero and Swistun 2009). Resulting forms of "organizational deceit" (Kroll-Smith, Brown, and Gunter 2000, 16) privilege the well-being of organizations over human health and deny agency to those who are most adversely affected by environmental threats. They allow governments and corporations to deny responsibility for alleged environmental health harm. One prominent strategy for doing so is to highlight the role of genes, diet, and lifestyle in disease causation over the role of pollutants, or stress that the latter cannot be blamed in isolation. This serves to maintain the legitimacy of both state and industry instead of questioning the unequal social order produced by industrial development and the profit imperative. It places the macroeconomic forces that subject some people to a disproportionate amount of harm beyond scrutiny.

Conversely, evidence required to prove environmental health harm and gain compensation from liable companies or state agencies is typically defined very narrowly (see Fortun 2001). A. Petryna's (2002) study of Chernobyl victims in post-Soviet Ukraine illustrates the politically driven, narrow legal criteria laid down by the state for what constituted radiation and attempts by citizens to gain compensation. Victims of radiation become "biological citizens" through their "demand for but selective access to a form of social welfare based on medical, scientific, and legal criteria that both acknowledge biological injury and compensate for it" (6). Their engagements with science and the state reveal that "the biology of citizens [is] a contested part of political processes" (21). Similarly, Waldman (2009) argues that the Indian state's "narrow definition of asbestos diseases enables it to officially document the lack of asbestos diseases experienced by Indian workers" (3). This, coupled with the delay between exposure and onset of disease, makes it "impossible for workers to be identified as victims" (16) of these ailments. Industries are therefore able to capitalize on competing epistemologies around the dangers of asbestos and claim to limit exposure to an acceptable level at which there is no evidence of harm (Waldman 2009, 21; Steingraber 2010).

These mechanisms of exclusion are supported by equally narrow, legalistic notions of justice for environmental health harm (Phillips 2012; Vanderlinden 2011). Conversely, victims are often accused of being emotional and lacking knowledge of the hazards or evidence that they are exposed to them. In this context, what may be deemed just by the justice system or the state may not be experienced as just according to victims' experiences and their moral codes. In fact, many studies of environmental justice show how people who raise the alarm and demand a cleaner environment are failed by the (justice) system and by the state, as well as by the companies (Das 2000; Fortun 2001; Phillips 2012). On this basis, Brown (2000) argues that it is necessary to ask "for *whose* standards, and by what version of *proof* is a standard of proof determined and employed?" (374). Steingraber (2010) similarly asks: "At what point does preliminary evidence of harm become definitive evidence of harm?" (9). The answer is not only scientific, but inevitably social, political, and economic.

This book illustrates how uncertainty about environmental health effects is articulated in the three case studies. It shows that, while powerful political and economic interests may maintain uncertainty (as much of the literature cited above points out), citizens do not always question these interests and might feel uncertain about environmental harm. They are caught between what they may recognize as relatively incontestable embodied evidence of pollution's harm and a feeling that their forms of evidence do not count. In some cases, their previous experiences with activism to stop pollution convince them that they are rather powerless to do so (Lora-Wainwright et al. 2012; Tilt 2010; Van Rooij 2006; Van Rooij et al. 2012). When citizens who live with pollution have also learned to take for granted the power relations underlying agnotology, and therefore these power imbalances are perpetuated rather than challenged, the complex subjectivity of those citizens requires closer attention. In turn, this might reveal some of the pitfalls of assuming that those suffering from pollution are likely to oppose it, question discourses of uncertainty, and compile evidence of harm.

### Environmental Health, Toxic Natures, and the Origins of Resigned Activism

Given the uncertainties surrounding environmental health and the powerful political and economic interests at stake, citizens can play an important role in questioning established standards for evaluating risk. The field of citizen science has raised important theoretical and analytical questions on the contestability of science, competing definitions of evidence, and citizens' role both in shaping knowledge and in mobilizing at local, national, and transnational scales.[5] "Popular epidemiology" presents a classic example of citizens' contribution in scientific controversies. While Petryna's biological citizens predominantly focus on demanding compensation based on parameters laid down by the state, popular epidemiology challenges existing scientific paradigms and science's claim to value neutrality (Brown, Kroll-Smith, and Gunter 2000, 18; see also Balshem 1993). First elaborated by sociologist Phil Brown, popular epidemiology refers to grassroots efforts begun by citizens, whereby they compile maps of local disease incidence, gather scientific data and other information, and also direct and marshal the knowledge and resources of experts in order to understand the epidemiology of disease (Brown 2007, 33–34; see also Brown, Morello-Frosch, and

Zavestoski 2012). Such citizen-science alliances took shape in the Woburn (Massachusetts) childhood leukemia cluster (Brown and Mikklesen 1997), and in parts of Louisiana's "cancer alley" where citizens teamed up with experts to map health effects onto the physical presence of pollution (Allen 2003; Lerner 2005; Ottinger 2013).

Studies of environmental health social movements are predictably concerned with the multiple forms of activism against pollution. Alongside citizen science, these include protests, resorting to the law and the media, lobbying polluting companies, and demanding governmental redress (Fortun 2001; Lerner 2005; Petryna 2002; Shevory 2007). They may take on relatively NIMBY (not-in-my-back-yard) connotations or be part of a more networked movement across different localities (Szasz 1994). But activism is by no means the only reaction. Just as often, communities may become resigned to pollution. Brown and Mikklesen (1997) argue that in Woburn victims fought back because children were affected, because they received media attention and help from legal and public health experts, and also because concern with toxic waste was on the rise nationally. However, it is much more common for victims of pollution not to fight back.

Such resignation has complex origins. When contamination is gradual and long-term—what Rob Nixon (2011) termed "slow violence"—victims may become accustomed to it and regard it as part of their lives (Auyero and Swistun 2009; Brown and Mikklesen 1997). I refer to this naturalization process as "toxic natures": in various ways across the three sites discussed in this book, pollution is part of the natural environment. Indeed, pollution is incorporated in the local environment, not only by virtue of being omnipresent—in the air, the water, the crops, locals' bodies—but also by affecting the very substance and appearance of the environment and its inhabitants. Baocun villagers, for instance, reported the appearance of a new kind of vegetable, which they termed "cabbage-turned-turnip." Due to the high levels of phosphorous in the soil, cabbages grew unusually long roots that resembled turnips. Similarly, claims that bodies can become accustomed to pollution (and that therefore, allegedly, they would be less affected by it) encapsulate locals' sense that contamination of their bodies was "normal," too common and self-evident to even comment on.

In these contexts, hazards may assume "the harmless aspect of the familiar" (Steingraber 2010, 8, quoting Carson 2000). Familiarity often goes hand in hand with a deep embedding of polluting activities within local

socio-political and economic contexts. On the most basic level, this boils
down to economic dependence on polluting firms. Erickson (1976) illustrated this in his study of the 1972 Buffalo Creek dam disaster in an Appalachian mining region. Given their complete dependence on the coal
company, residents behave "*as if* they trusted the experts because it would
have been socially and psychologically unviable to do anything else"
(Wynne 1996, 52).

But the relationship between polluters and residents is often more complex than simple economics. Residents invest emotional and financial
resources in the locality, as was the case of residents of Legler, a suburb of
New York built on a municipal landfill. "Caught up in their families, work,
home, and the other concerns of the American lifestyle," contamination
was "virtually inconceivable" (Edelstein 2004, 29). Similarly, despite their
knowledge of pollution, residents of the Argentinian shantytown of Flammable have become slowly tied to the locality, taking roots in the neighborhood through work, family, and friendship networks, and therefore they
play down the dangers they face (Auyero and Swistun 2009, 86; see also
Brown and Mikkelsen 1997, 62).

Under such circumstances, communities affected by pollution may prefer to deny its existence or its gravity. Indeed, across my three case studies,
widespread ailments correlated with environmental contamination were
not mentioned as ailments at all unless research participants were prodded
by the researcher. Denial serves as a strategy, a psychological defense against
contamination (Brown and Mikklesen 1997, 53). By contrast, to identify
pollution's harm would also imply the recognition that government and
industry did not protect residents' welfare, and the explicit acknowledgment of their own powerlessness, neglect, and marginality (Brown and
Mikklesen 1997; Erickson 1976; Wynne 1996).

Pollution has deep social effects too: it may create a sense of social mal-
aise, instability, anxiety, depression, anomie manifested as exploitation,
selfishness, and loss of confidence in commercial practice, government,
and science (Williams 1998, 14–15, see also Auyero and Swistun 2009;
Brown and Mikklesen 1997; Edelstein 2004; Singer 2011). The community's
"lifescape" (their entire world and their experiences) is radically disrupted:
people experience a loss of control; they learn to see the environment as
uncertain and harmful. The home, which is usually regarded as a safe
haven, becomes a site of risk and fear (Edelstein 2004). The experience of

contamination may convince residents that there is "no safe place" where they could seek refuge, that pollution is pervasive and inescapable (Brown and Mikklesen 1997). This experience in turn often divides communities and causes social conflict; activists are ostracized, stigmatized, and portrayed as oddballs intent on undermining social stability (Brown and Mikkelsen 1997; Kirby 2011).

This literature is useful in posing questions on citizens' role in environmental health contestations in China. Chinese citizens—particularly among the urban middle classes (see Johnson 2013a and b), but also some residents of "cancer villages"—have increasingly contested environmental pollution through instances of activism, resorting to new technologies, social media, and environmental NGOs. Chapter 2 will illustrate how residents of some self-identified "cancer villages" have attempted to gain attention, legitimize their claims, and obtain redress, and the challenges they face in doing so. However, rural environmental activism remains largely invisible, silenced, and ineffective as a measure to significantly decrease pollution. Partly, the absence of a social movement premised on pollution's effects on health is due to the inherent complexity of environmental health and the difficulties in gathering scientific evidence that particular ailments can be conclusively traced to exposure to particular chemicals. Uneven access to evidence and resources to question existing claims are a challenge for any environmental justice movement (Steingraber 2010), but they are all the more haunting in rural China, where resources are limited and the government largely holds the monopoly on tests and their results. Even when tests are carried out, either the results are not shared, or their reliability is deeply doubted by the local population.

Yet the absence of an environmental health social movement stems not only from the nature of the hazard but also from the nature of society (Kasperson and Kasperson 2005). Indeed, a number of recent qualitative social science analyses of environment and health in China have begun to highlight the importance of social and economic dependence and wider opportunity structures in shaping the local population's perceptions of and responses to pollution (Chen et al. 2013; Deng and Yang 2013; Lora-Wainwright, Zhang, Wu, and Van Rooij 2012; Tilt 2006, 2010; Van Rooij 2006; Yang 2010b). Building on these studies, I argue that knowledge of pollution cannot be separated from the many other challenges locals face—such as finding work, paying for healthcare, and improving their family homes. As industry becomes enmeshed with the local community and locals become more and more closely tied to their place of residence, pollution can come to be regarded as a fact of life, and only one of a number of potential causes of illness. Conversely, common illnesses potentially correlated with pollution (such as nose bleeds in Baocun) become regarded as "normal." As a consequence, responses to pollution also become embedded within social, economic, and political relationships, and the everyday challenges of making a living and attaining a good life. This book is devoted to elucidating these processes.

Across the three main case studies, dependence on polluting activities, social divisions generated or exacerbated by such activities, (limited) opportunity structures, and past experiences with seeking redress reinforce locals' sense of uncertainty in attributing illness to pollution and weaken their sense of entitlement to a healthy, clean environment. Uncertainty in turn further crystallizes these social, economic, and political configurations by undermining villagers' ability and willingness to claim they are physically harmed by pollution and to act collectively on this basis. I use the term "resigned activism" to highlight these powerful, if at times counterintuitive, dynamics among perceptions, attitudes, and responses to pollution. Rather than assume a unidirectional relationship among these elements, I argue that the results of previous activist efforts contribute to forming new perceptions and attitudes to pollution. Perceptions of and responses to pollution in other words are part of a shifting, complex entity, which in turn is affected by its constantly changing social, economic, and political contexts. Resigned activism then refers to efforts villagers undertake routinely, individually or as a group, to counter or avoid pollution. But it also refers to the simultaneous processes through which pollution comes to be regarded as a normal and unavoidable part of the natural environment. It attends to the effects that subtle forms of environmental engagement may have not only on the environment but also on the subjects themselves.

The expression "resigned activism" is intended not only to convey the mutual relationship between perceptions and actions, but also to bridge the wide spectrum of responses that are usually regarded to be separate and even incompatible. It includes well-oiled activist strategies, such as petitions, blockades, and protests, alongside lifestyle choices (drinking tap water and closing windows at night), requests for scientific tests on water and crops, and demands for piped drinking water. Elements in this complex spectrum of attitudes and responses—from various forms of opposition to the denial of pollution's harm—may be visible at different times among affected communities. Examining only certain forms of activism would fail to capture the much more uneven nature of engagements with pollution. By contrast, resigned activism encapsulates the development of attitudes and responses to pollution over a longer time span, to include both highs and lows in cycles of activism, as well as the hesitation, uncertainty, despondence, and acquiescence that pervade much of these processes.[6] Full-length monographs on single case studies of resistance to pollution have shown aptly how communities' understanding of pollution, their strategies, and their relationship to the polluters, state agents, and various stakeholders change over time.[7] This book examines these dynamics in the Chinese case. It maps how and why resilience in opposing pollution may coexist with resignation to it. This in turn requires a careful redefinition of both resistance and resignation, as well as their potential overlaps.

## China's Environmental Challenges and Environmentalism

### The Basic Challenges

China's environmental problems are multifarious and therefore difficult to summarize. They may take the form of both pollution accidents and of routine, widespread pollution. They are the result of a combination of forces including rising affluence, globalization of manufacturing, urbanization, and climate change (Shapiro 2012). For instance, the exponential growth of waste in recent years creates challenges for waste disposal that in turn materialize as potential threats to the environment: dumpsites may pollute the soil and water, incineration may cause air pollution. Electronic waste is potentially poisonous, when part of the recycling process leeches harmful heavy metals into the environment (see chapter 5). Likewise, the huge expansion of industry, and resource extraction and processing, have several effects on the environment, including air, soil, and water pollution from heavy metals and toxic organic chemicals—which in turn may directly affect food safety, as well as the health of those who live in the vicinity of industries (see chapters 3 and 4). Exposure to pollutants in food—whether they are heavy metals from nearby industries, excessive farm chemicals, or noxious substances willfully added during food production (as was the case for the melamine-contaminated milk powder)—poses additional risks to human health. Even measures that were heralded as the key to sustaining development in the aftermath of the global economic crisis, such as the drive to encourage consumption and urbanization, also have huge repercussions for the environment. These problems may be exacerbated when they intersect with particular local geographical conditions such as climate, lifestyles, and patterns of production and consumption. For instance, water scarcity may increase the impact of water pollution by making local residents more dependent on polluted water sources (see chapter 4).

China's particular pattern of development and the speed of that development mean that it is concomitantly faced with "'traditional' environmental impacts on health associated with poverty; an increase in 'transitional diseases' related to rapid industrialization and urbanization; and what are known as 'the diseases of affluence,' all at the same time" (Holdaway 2013, 257). Much literature to date focuses on health problems related to industrialization and urbanization. A 2007 collaboration between the World Bank and China's State Environmental Protection Agency (or SEPA, since 2008 known as the Ministry of Environment) to assess the cost of pollution in China identified cancer as the main cause of death in China (World Bank and SEPA 2007). This study also showed that mortality rates for cancers associated with water pollution, such as liver and stomach cancer, are well above the world average. According to unreleased World Bank statistics, 750,000 people die prematurely in China each year due to high pollution levels (McGregor 2007). More recently, an international study using long-term data found that China's air pollution has cut life expectancy by an average of 5.5 years (Hook 2013). A 2015 documentary by former investigative journalist Chai Jing, *Under the Dome*, exposed China's pollution, its adverse effects, complex origins, and uneven solutions through field investigations and interviews with scientists and officials. The huge response it received—both in terms of the millions of hits it received within hours of its release and of the swift crackdown that quickly followed—is evidence of widespread concern about pollution's effects, among the general population and the upper echelons of the government alike.

Environmental health problems are posed not only by the speed of China's development but also by some trends that are relatively peculiar to this national context. First, much of China's industry, unlike in many other parts of the world, has been historically located in the countryside and has involved a rapid turnover in the type and scale of industry (Holdaway and Wang 2013). Many of these industries later closed or changed product lines or forms of ownership, meaning that rural China has been affected by pollution from different sources and that liability is often unclear (ibid.). Migration adds a further level of complication: rural residents otherwise dependent on agriculture have often moved in search of waged work, including in industry and other polluting sectors, but their very movement complicates any effort at tracking how their health has been affected by working in these businesses (ibid.; Holdaway 2014).

The geographies of development and pollution are regionally and locally uneven. Indeed, under growing scrutiny and pressure to clean up, many polluting firms have opted instead to move to poorer regions and further into the countryside. As a consequence, many rural residents either do not benefit directly from development or do so at great cost to their health. "Cancer villages" (clusters of high cancer incidence typically correlated with pollution; see chapter 2) are one of the most notorious side effects of such rampant rural development and a classic example of "transitional diseases" (Holdaway 2013, 257). These inequalities are compounded by the continued disparity in welfare provision between rural and urban areas. Welfare provision in China's countryside has doubtlessly improved after the introduction of rural healthcare cooperatives in recent years, but it remains limited (Yip et al. 2012). Efforts to carry out "urban and rural integrated planning" (Chen and Gao 2011) might mitigate these disparities, but they might also result in bracketing rural areas for industrial development and therefore affecting rural populations disproportionately.

The movement of polluting projects into rural China is coupled with another trend, whereby those with more financial means move away from pollution, buy bottled water, and install air purifiers. Most rural Chinese cannot afford to do this. Unless strict environmental protection policies are implemented, inequalities in the distribution of pollution and harm are only likely to increase as industrialization and urbanization continue to move west and as tech-savvy, wealthier communities demand a cleaner environment. China's role as the world's factory exacerbates these trends, as additional opportunities for regional development may transform into further pressures to industrialize at all costs, to emulate and catch up with more developed coastal regions.

### The Historical Evolution of China's Environmental Problems

The origins of China's environmental challenges are complex, and not only limited to the contemporary period (Elvin 2004; Shapiro 2001). However, the past four decades have witnessed particularly severe environmental degradation. Following the death of Mao in 1976, China embarked on a massive program of social and economic reforms. Rural industrialization is one of the key features of the reform period (Oi 1999). Township and Village Enterprises (TVEs) established during the reform period soon became a vital and unregulated engine for local development. The partial liberalization and privatization of resource extraction, which resulted in the opening of township and village mines and small private mines, also led to a further expansion and deregulation in these areas (Wright 2011). Low capitalization was crucial to the economic success of TVEs and private mines. Profit considerations determined investment in industrial infrastructure and methods of mining and processing, with little regard for longer-term environmental impacts (Tilt 2010; Wright 2011). The combination of high-potential economic benefits and severe and acute effects on environment and health makes pollution from mining and industrialization one of the most prevalent causes of contention and unrest among rural residents.

The grave environmental consequences of this development model soon became apparent, and the early reform mindset—"pollute now, clean up later"—has come under growing scrutiny. Since the late 1970s, China's government has developed an impressive body of environmental protection policies and legislation. But effective local implementation has remained uneven. Several studies highlight challenges to environmental protection posed by governance mechanisms and the legal system. Many of these are by now well-known: poor state environmental agencies' capacity to monitor and enforce compliance; lack of capacity among local governments; the decentralization of environmental enforcement to local officials with conflicting interests, particularly a tension between environmental protection and economic targets; lack of coordination and an ambiguous responsibility structure divided over a number of ministries and other agents; poorly designed policy instruments; and general, vague, and aspirational legislation that often falls short of being locally feasible or carrying a strong enough disincentive to pollute (see Carter and Mol 2007; Economy 2004; Kostka and Mol 2013; Tilt 2010; Van Rooij 2006).[8]

Changes put forward in the past decade are intended to tackle this notorious implementation gap. In 2008, the bureaucratic rank of the State Environmental Protection Administration, SEPA, was raised to that of a full Ministry of Environmental Protection (MEP), endowed with better capacity to enforce compliance. Five "Regional Supervision Centers" have been established to oversee local implementation and coordinate trans-provincial environmental disputes. Growing awareness of the threat that pollution presents to health is evident in the emergence of "environment and health" as a field of research and policy (Holdaway 2010, 2013). Among the new administrative measures and tools that have been intro-duced to enhance compliance are: mandatory 'binding' environmental targets for government; private enterprises adopting targets for energy effi-ciency and carbon emission; the strengthening of environmental courts; experiments with market-based instruments to reduce emissions and set electricity pricing; payments for environmental services (PES); and the creation of new decision-making structures, such as interdepartmental committees to coordinate implementation (Kostka and Mol 2013; see *People's Daily* 2015).

China's former leaders, Hu Jintao and Wen Jiabao, professed a commit-ment to "sustainable" "scientific development," harbored within a "harmo-nious society," and the current leadership emphasized the need to build an "ecological civilization" as part of the 12th Five-Year Plan announced in late 2012. Yet implementation and public participation still remain serious obstacles, and local interests continue to dictate development agendas (Kostka and Mol 2013). National polices are only implemented when locali-ties are under direct and constant attention. Environmental protection is only one of a number of competing local priorities. Economic performance is still a vital target in the formal performance evaluation of cadres (Edin 2003; Ho 1994; Whiting 2001). More importantly, clashing targets put for-ward by the central state itself pose a challenge to local cadres' ability to implement central directives. A National Plan for Environment and Health Work, designed to be implemented at the county level, has made limited progress due to the lack of financial and technical resources in many locali-ties (Holdaway and Wang 2013).

Innovations proposed by the new leadership of Li Keqiang and Xi Jinping might offer some room for hope. In recognition of the gravity of the situation, current premier Li Keqiang declared "a war on pollution" at the opening of the 2014 annual session of the National People's Congress (Branigan 2014). The continued prominence of the concept of "ecological civilization" in recent key state documents (*People's Daily* 2015) entails comprehensive and detailed plans (including clearer standards, mechanisms, and a new lifelong evaluation for officials) that suggest the Chinese government is taking environmental protection increasingly seriously (see Geall 2015).

Only time will tell whether the most recent innovations may bring the desired effects. Challenges are likely to remain in historically poor regions with little revenue alternatives to polluting firms, and where therefore local governments depend on such firms to raise tax revenue necessary to support public services and villagers rely heavily on them for employment (Tilt 2010; Wright 2011). Many of these areas (albeit not all) are in western China, which is historically less developed than its eastern seaboard. There, mining and industrialization brought increasing employment opportunities. As Brian Tilt's 2010 ethnography of one of these regions (Futian, in Sichuan) shows, local officials and communities alike face genuine challenges. Township and County Environmental Protection Bureaus "must weigh the ecological and health consequences of industrial pollution against the economic and fiscal benefits of industrial production" (Tilt 2010, 112). This engenders a phenomenon known as "local protectionism," whereby local governments protect polluting firms, with the complicity of the local population (Van Rooij 2006). This is particularly the case when they depend on polluting firms for employment (Chen et al. 2013; Deng and Yang 2013; Lora-Wainwright, Zhang, Wu, and Van Rooij 2012; Tilt 2010; Van Rooij 2006). This trend raises important questions about the potential for citizens to become agents of environmental protection. The current book explores their complex positions.

## Key Actors and Features of a Growing Environmental Movement: Rural China in Comparison

Given the gravity of China's environmental challenges, the limits in the government's capacity to enforce environmental regulation, and the vicious cycle of low compliance and weak enforcement, civil society has been a vital complement to top-down forces (Economy 2004; Geall 2013; Shapiro 2012; Zhang and Barr 2013). In the last decade, the Chinese state has become significantly more responsive to societal pressures. In part, this is due to the fragmentation of authoritarian rule by local diversities in the modes and degrees of implementation of laws and regulations and to the emergence of a new and broader set of "policy entrepreneurs" (Mertha 2010). Analyses of environmentalism in contemporary China have grown in recent years, reflecting its increased prominence. In this section I will outline the range of strategies and actors that may take part in environmental action. Forms of activism may involve: litigation, petitions, direct negotiation with relevant firms, appeals to administrative regulators, appeals to the media, the Internet, environmental NGOs, advocacy networks and elite allies, and citizen protests to put pressure on firms or on administrative authorities. Each of them brings its own potential benefits and liabilities.

**Direct Negotiation, Petitions, and Rightful Resistance**   Original fieldwork in Baocun and Qiancun, and a survey of media reports and academic publications containing "cancer village" or "environmental politics" in the keywords on CNKI (China National Knowledge Infrastructure, the main database of Chinese newspapers and academic publications), suggest that direct negotiation with polluting firms and with local government are among the most widespread strategies adopted by villagers, particularly in the early stages of contention (see Van Rooij 2010). This is often done alongside petitions to the local government in an effort to raise the alarm and obtain redress (Brettell 2003). The Internet, and microblogs in particular, play a key role by helping to raise concern among the wider population and organize actions of this kind (Yang 2011, 2013).

Indirect reference to laws and regulations in these negotiations is a common strategy to present their demands as lawful. The case of Qiugang, a "cancer village" in Anhui province portrayed in the 2010 documentary *The Warriors of Qiugang,* is a pertinent example. Qiugang villagers' ability to ground their grievances in the law and to quote official speeches by Hu Jintao proved to be powerful weapons, even when litigation itself had failed them (Wang 2011). This strategy is known widely as "rightful resistance" (O'Brien and Li 2006; O'Brien 2013)—that is, citizens' resort to existing rules and regulations to present their resistance as "rightful." However, attempts to approach higher levels for support because they are seen as more trustworthy may not always have positive outcomes, and demands that escalate to higher levels are also likely to be more complex to resolve

(Michelson 2008). In some cases, what citizens strive to present as rightful resistance is strictly speaking unlawful. For instance, protesting against projects that have passed the Environmental Impact Assessment (EIA) is actually illegal and places both local governments and citizens in difficult positions (Hu and Tilt 2012). For their part, citizens have their reasons to doubt that EIAs have been done properly. While EIAs ought to be carried out for all major projects and should be openly available to the public, gaining access to such information is very time consuming and bureaucratic, and, more often than not, the data is actually unavailable. In some cases, the information contained in the EIA is simply wrong (Hu and Tilt 2012; Johnson 2013a). Only rarely have villagers succeeded in accessing such information.

**Litigation**   Legal possibilities to obtain compensation for pollution have grown, and, as a consequence, so have environmental litigation cases, though they remain very low compared to petitions (Stern 2013). Yet knowledge of a risk does not always translate into action, and most action remains beyond formal institutions (Diamant et al. 2005). Surveys of litigation cases suggest that disputes emerge when aggrieved citizens forge a group identity; they require social and financial resources, support structures (especially from the local state and intermediary institutions), community solidarity, coalitions (including assistance from experts, NGOs, and the media), and leaders capable of gaining support (Diamant et al. 2005; Van Rooij 2010). Even when citizens resort to litigation, they still face many obstacles ranging from ability to afford legal assistance, to the challenges of providing evidence of damages, demonstrating the existence of a polluting act, and the liability of a given industry (Van Rooij 2010, 68–69). Although civil environmental litigation is occasionally successful, it is a relatively weak tool for environmental protection (Stern 2011, 310; Stern 2013), particularly in rural areas. In Qiugang village, local resident-turned-activist Zhang Gongli brought several unsuccessful lawsuits against local polluters, despite seemingly clear violations of the law. Lawsuits also typically focus on "post-hoc monetary compensation," and therefore serve more as an alerting mechanism for extreme abuses than to prevent pollution or demand remediation (Stern 2013). They tend to be a relatively rare strategy, a last port of call when other avenues have failed. Their success also depends heavily on what happens outside the court, particularly on the presence of populist pressure, sometimes galvanized by the media (Stern 2013).

**Media Coverage**   Appealing to the media is another potentially crucial strategy (Geall 2013; Yang 2011). Y. Cai (2010) has shown that media coverage is a key determinant of successful collective action. Typically, if a case becomes well known, the central government is more likely to intervene and put pressure on the local government to address locals' complaints. Most recently, the campaign to demand that PM 2.5 (particulate matter smaller than 2.5 micrometers and therefore most harmful to health) in Beijing's air be monitored and results released was successful largely because of media attention (Fedorenko and Sun 2015). Reports can help to tip public opinion in favor of a movement. But when polluting firms play a vital role in the local economy, the potential of media attention to aid regulation and closure of polluting firms is limited (Tilt 2010; Tilt and Xiao 2010). Media attention is also hard to gain, and journalists are constrained in the ways in which they can frame their reports. The national media can generally be freer to report cases that do not implicate officials above the county level (Stern 2011, 306), but with potentially more controversial and broader problems they need to choose their focus carefully. At any rate, media coverage does not automatically result in redress, and in many cases the pressure dissipates once media attention falters (Chen et al. 2013).

**NGOs**   Contacting or establishing NGOs may serve a vital role, often alongside the media (see for instance: Fürst 2012, 2016; Johnson 2010; Morton 2005; Spires 2011; Steinhardt and Wu 2015; Wu 2013; Yang 2005). Studies of environmentalism in China have predominantly focused on urban environmental NGOs, paying much less attention to grassroots forms of activism. This is understandable, given that the emergence of ENGOs (Environmental NGOs) in the mid-1990s is widely regarded as the starting point for China's environmental movements. The environmental field has the largest number of NGOs and longest history of public policy advocacy (Hildebrandt 2013). Environmental NGOs have grown exponentially in the past twenty years, and according to Wu Fengshi (2013) they are showing a certain degree of maturity, with a presence beyond large urban centers and in more regional areas. ENGOs have evolved to become very diverse, ranging from well-established registered NGOs like Friends of Nature, the Institute of Public and Environmental Affairs (IPE), and the Center for Legal Assistance to Pollution Victims (CLAPV), to more grass-roots organizations like Green Beagle and Nature University, both founded by journalist-turned-activist Feng Yongfeng. Whereas in their early phase of development, ENGOs relied on a few leading activists, scientists, and jour-nalists, in the aftermath of the 2008 Sichuan earthquake, a younger, less recognizable, and more diverse generation of social entrepreneurs and NGO leaders has emerged and spread far beyond Beijing (Wu 2013; see also Geall 2013 and Boyd 2013 for a genealogy of recent campaigns). Civil society however shows signs of tightening under the leadership of Xi Jinping since 2012. A new Overseas NGO management law approved in the spring of 2016, and coming into effect in 2017, mandates that any group wishing to operate in China must register with public security officials and poses many restrictions on their activities. This will affect not only foreign NGOs, but also domestic NGOs relying on foreign funding, and affect the issues on which they are able to focus.

Like the media, ENGOs need to be strategic about the issues on which they focus, and the manner in which they frame them, particularly when they are large and influential. The campaign against dams on the Nu River amply exemplified the importance of choosing a relatively noncontrover-sial focus that drew on global discourses of biodiversity. Such focus on protecting biodiversity and endangered species, rather than on the welfare of the local population and compensation for lost homes and land, was instrumental to the campaign's success in (temporarily) halting dams (Litzinger 2007, 289–91; Mertha 2010; Yang and Calhoun 2007). In this context, Peter Ho (2008) has referred to China's ENGOs as "embedded," because they operate closely with state structures and avoid direct opposi-tion. Unlike NIMBY (not-in-my-back-yard) movements, which are more contentious, China's ENGOs are typically localized, loose, and informal, mostly embrace a rules-based approach to public participation and engage with educational activities, tree planting, or waste collection (Johnson 2010). They are "the pivotal organisational basis for the production and circulation of greenspeak" (Yang and Calhoun 2007, 213). "Greenspeak" draws on the conceptual space of donors (often led by the mainstream global discourse of sustainable development) and of the Chinese govern-ment (therefore avoiding openly political issues) (Yang and Calhoun 2007,

214). In addition, the success of a campaign depends heavily on the ability of greenspeak agents to raise concern in the public sphere. Mertha (2010) refers to these agents as "policy entrepreneurs"—including experts, disgruntled officials, journalists, and NGO campaigners—who are able to mobilize broad-based coalitions. His study highlights the importance of networks and coalitions, rather than NGOs in isolation (see also Wells-Dang 2012). In this context, Hathaway's concept of "environmental winds" (2013) provides a fluid model for understanding the flows of environmental ideals within and beyond China and ways in which to think of entanglements between state agents, NGOs, scientists, villagers, and Chinese and foreign conservationists.

Predictably, registered NGOs that focus on regulating pollution are relatively few, due to the sensitivity of the topic, lack of relevant training among NGO staff, and difficulties in securing funding for this focus (Fürst 2016). In her detailed 2016 study of China's environmental NGOs and their efforts and achievements in regulating pollution, Fürst estimated that out of 8000 registered environmental NGOs, only between 150 and 200 focus on pollution regulation, and their efforts are fragmented and weak. Her research suggests that these NGOs predominantly regulate "through leverage," in other words, they rely closely on support by state agencies, international companies, the media, the market, and public opinion to influence the behavior of their targets. Their effectiveness is affected not only by underlying state-civil society relations in China and by the sensitivity of particular issues, but also by the level of scientific complexity involved, by the nature of this particular policy domain, and the structure of the policy networks (Fürst 2016, 376 and chapter 12).

**Community Advocacy and Crowd-Sourcing**  Advocacy networks occasionally include community advocacy, though this is still relatively rare in China (Wells-Dang 2012). Such networks tend to be among elite, urban-based allies (Johnson 2013a and b). We hear less about the participation or viewpoints of the communities directly affected, whose perspectives may differ substantially from those of urban elites and advocacy networks (Litzinger 2007). Recently, there have been some signs of change, both in terms of NGOs' embeddedness and avoidance of conflict, and in terms of their joining forces with local communities. In the past, NGOs have often been criticized for choosing soft targets and moderate strategies, rather than making a stand in major cases such as the pollution of the Songhua River in November 2005 caused by explosions at a petrochemical plant in Jilin, or the blast and oil spill in Dalian, northeast China, in the summer of 2010 (Tang 2012). More recently, civil society organizations have become more confrontational and increasingly active in supporting citizen action against pollution (Fürst 2016). For instance, two NGOs— Friends of Nature and Chongqing Green Volunteers Union—supported by CLAPV joined the Qujing Environmental Protection Bureau to sue Luliang Chemical Industry for dumping extremely harmful hexavalent chromium in rural Qujing. Regardless of the outcome, this is a breakthrough case for NGOs to act as plaintiffs in environmental public interest litigation (CLAPV 2012).

Relatively new and grassroots NGOs also work more closely with the public to identify issues of concern, rather than imposing urban and middle-class environmentalist metanarratives upon local populations, as had largely been the case in the campaign against dams on the Nu river (Litzinger 2007). One significant development is efforts of these new and grassroots organizations to join forces with local populations in order to gather data on pollution. For instance, grassroots NGOs Green Beagle and Nature University provided portable handheld detectors to measure air quality (Zhang and Barr 2013). But these efforts also extend beyond air pollution. Former journalist and Goldman Prize winner Ma Jun established the Institute of Public and Environmental Affairs (IPE), which for several years has compiled online pollution maps to inform citizens about environmental hazards, demand transparency, and put pressure on polluters to clean up. Similarly, newspaper photographer Huo Daishan played a key role in raising the alarm about pollution in the Huai river basin. In 2000, he formed a group called Guardians of the Huai River, which trained hundreds of volunteers who now work in teams to regularly monitor the river and conduct water testing, pushing companies to implement pollution-control measures (Baike.com 2016). More recently, a nonprofit group called the IT Engineers for Environmental Protection Association created "danger maps," an open platform that resorts to crowd-mapping and allows users and NGOs to upload their own pollution data. These efforts at documenting pollution, which are gradually spreading into rural China, are powerful signs of change. However, just as several of the strategies outlined above, they remain limited to a minority of cases.

**Urban Protests**   Frustrated by the constraints and limited effectiveness of framing complaints within parameters set by the state and by the lack of effectiveness of attempts to negotiate directly with firms or to appeal to administrative regulations, citizens have engaged in less institutionalized contentious action (Deng and Yang 2013, Lora-Wainwright et al. 2012). This is the most daring strategy. Significantly, even when contentious action is involved, protestors endeavor to frame their activities in such a way as to portray themselves as rational, concerned citizens and avoid confrontational connotations (Johnson 2013a; A. Zhang 2014). In 2007, a groundbreaking protest in Xiamen (Fujian), against a paraxylene (PX) plant, succeeded in halting plans for the project (Ansfield 2013). As large protests are illegal, organizers described their activities as a "stroll." The following year, similar protests took place in Shanghai against the construction of a magnetic levitation train line.

Arguably, these kinds of actions have intensified in the past few years.[9] Citizens protested against a PX plant in Dalian in 2011. The year 2012 alone saw three protests of this kind, which were widely covered in the international media. In early July, more than 10,000 residents of Shifang (Sichuan) clashed with the police and succeeded in halting a planned molybdenum copper refinery (Hook 2012). Barely weeks later, similar protests erupted in Qidong (Jiangsu), as thousands occupied the local government building, filled the streets, overturned at least one car, and suspended plans for a wastewater pipeline (Tejada 2012). In October, demonstrations in Ningbo over the planned expansion of a petrochemical state-run Sinopec plant (already one of the nation's largest refineries) led to scrapping a controversial PX facility, and to promises of public consultations and greater transparency (*South China Morning Post* 2012). More protests against a PX plant took place in Guangdong in 2014 (Gu 2016; Lee and Ho 2014).

These are examples of a rising number of citizen demonstrations against existing pollution or plans for what are understood to be polluting projects that local residents fear would compromise their environment and health (Tang 2012). Protests are never the first port of call, but rather a desperate cry for attention. In Ningbo, for instance, protests only took place after authorities ignored weeks of petitioning. Undeniably, protests put pressure on local governments to clean up their act, but promises made may be forgotten as soon as the crowd disperses. Microblogging and web-based chat groups can serve as platforms for public debates and to organize these forms of action (Yang 2011, 2013). However, attempts to present protests as nonconfrontational may not always succeed, and the line between peaceful strolling and protests is blurred at best. Indeed, the dangers to protesters are considerable. Arguably, the higher the profile of protests and their attracting international coverage, the higher are their chances of obtaining redress, but also the higher the political risks to those involved in protests.

Behind these protests and behind much of environmental activism portrayed in the media is a demographic that officials cannot ignore—rising numbers of urban, educated, middle-class professionals with economic, and, increasingly, political clout who form the backbone of the new affluent China (*South China Morning Post* 2012). These protests largely employ the language of personal rights and protecting one's home (Su and Link 2013), and strategically present their resistance as "rational," in contrast to the irrational behavior of polluters and colluding local governments (Johnson 2013a; A. Zhang 2014). While urban homeowners' protests are more prominent in the media, rural complaints about pollution have been less visible (see also Yang 2010a). Activism in "cancer villages" represents an exception to the general lack of visibility of rural environmentalism, but these cases are by no means representative of how villagers in severely polluted areas react to such pollution. Often, these cases suggest that citizens inherently oppose pollution and the only real obstacle to them doing so more effectively is their limited ability to participate in the political process. This book challenges these assumptions. The range of options for activism outlined in this section offers a useful framework, but ultimately only qualitative case studies can illuminate how, when, and why these options are operationalized in practice and how they intersect with a pervasive sense of resignation. The latter deserves just as much attention if we are to grasp the complex, shifting, and diverse experiences of those living with pollution.

## Environmental Consciousness and Activism in Rural China

The predominant focus in current research on the role of NGOs, the media, and litigation as potential forces in aiding environmental protection implementation leaves out much of citizens' actions, which often take place beyond these means and rest more firmly on small protest and direct appeals to local officials and polluting firms. Indeed, countless villages are suffering from pollution without being able to gain attention, let alone redress. The processes underlying their decisions to take particular kinds of actions remain little understood. How do concerns with pollution crystallize and how does rural environmentalism emerge? While it is widely agreed that citizens may play an important role in pollution regulation, very little is known about the intricate processes through which villagers themselves understand environmental health threats (Tilt 2010; Weller 2006). How do they establish a link between a given threat (for instance, contaminated food or air pollution) and their experiences of illness? What do they consider to be evidence of environmental harm to health? Who is involved in these contestations and what are their consequences?

Studies of environmental consciousness have tended to focus on the urban middle classes. According to the post-materialist thesis, concerns for the environment only arise among those who are not preoccupied with meeting their basic subsistence needs. The poor, in other words, cannot afford to care. Writing against this view, Robert Weller argues that the rural Chinese he studied *are* "concerned about environmental effects on the health of their children and the quality of their crops. ... They lack environmental consciousness only in the sense that they are not concerned with the same issues as national and global elites, or as people who write questionnaires about values" (Weller 2006, 57). "The bulk of environmental action," writes Weller, takes place "among people whose motivations are above all local and personal" (133). With this in mind, he argues that it is at least inaccurate to accuse people of lacking an environmental consciousness if the parameters used to assess it are different from those used by local people themselves.[10] In this same spirit, the current book places attention on local parameters of environment, health, and development, and how they evolve.

Research on the formation of environmental consciousness among villagers and on their responses to pollution is still limited. A number of recent articles have examined the social, political, and economic context in which local communities understand and respond to pollution. They have shown that relative dependency on polluting firms and the relationship between villagers and polluters mold their attitudes to pollution. Different occupational groups react differently to pollution, and, predictably, those whose livelihood depends directly on industries are least likely to complain (Tilt 2006). Deng Yanhua and Yang Guobin (2013) argued that villagers in Zhejiang opposed the siting of a polluting industrial park in their vicinity, but were more accepting toward equally polluting plastic recycling operations started later by other locals. Research by Chen Ajiang and his team (Chen et al. 2013) finds insider-outsider interactions shaping responses to pollution in the context of their research on "cancer villages." Both Tilt (2013) and Lora-Wainwright (2013d) found that, albeit in different ways, residents of polluted villages take a relatively individualistic approach to coping with pollution, given its seeming inevitability and uncertainties about its effects.

The only full-length studies to focus closely on rural China and on villagers' experiences of pollution are by Tilt (2010) and by Chen and his team (Chen et al. 2013). Research by Chen and his team (some of which will be examined in detail in chapter 2) is based on years of fieldwork in heavily polluted areas, including a number of "cancer villages." Their book covers a range of sites through in-depth case studies that highlight the uneven understanding of pollution among villagers, equally uneven evidence of a correlation between pollution and health effects, and the diverse reactions of local communities to pollution. They attribute such diversity to different local political economies, different relationships between communities and polluters, and varying levels of interactions with outside actors, such as the media, lawyers, and higher government authorities. Bryan Tilt's (2010) study of sustainable development in an area of rural Sichuan reliant on metal industry shows that pollution, for rural dwellers, has become a fact of life. But this is not taken as simplistic evidence that locals have become resigned to pollution and accept it because they are too poor to care. To the contrary, Tilt argues that people engage critically with the need for development and that pollution is experienced in deeply local ways. Environmental issues "are embedded in specific cultural, economic, and political formations and ... must be studied *within* those formations" (2010, 153).

Building upon Weller's understanding of environmental consciousness and on Tilt's work, this book examines the origins and implications of the parameters of environmental health consciousness that emerged in some severely polluted sites. It argues that these very parameters are inseparable from local power relations; they affect and are affected by various activist practices embraced. It shows that the link between concerns and action is not linear or automatic. Indeed, not taking action is a common way for villagers to deal with injustice given the tremendous obstacles they face in taking successful action, even when they may be acutely aware of being victims of harm (see Michelson 2007; Tilt 2010). In view of this, the book zooms out on the longer processes of knowledge formation around pollution and looks beyond single strategies of contention to encapsulate the longer and more complex development of concerns with pollution and strategies to deal with it. It approaches forms of activism as embedded in the habits and values of village life (see Chen et al. 2013).

Ultimately, only a close analysis of particular cases may shed light on the vicissitudes of local activism and resignation to pollution and their complex dynamics. This requires focusing attention on: the evolving perceptions of pollution among villagers, polluters, and the local government; how the relationship between these intersecting groups shifts over time; and the uneven patterns of dependence, gain, and suffering. It also requires that we disaggregate the category of 'villagers' to better understand how engagements with pollution may vary across age, gender, occupation, economic standing, and other lines. This in turn enables an account of how villagers' identities and positions were shaped by the presence of pollution.

## 2 China's "Cancer Villages": The Social, Political, and Economic Contexts of Pollution

### Introduction

In 2013, the Chinese government publicly acknowledged the existence of "cancer villages" (clusters of high cancer incidence typically correlated with pollution), feeding a controversy that started in 2001 with the first appearance of the term.[1] An official document titled "Guard against and control risks presented by chemicals to the environment during the 12th Five-Year Plan (2011–2015)" (Ministry of Environmental Protection 2013a, 10) stated that the widespread production and consumption of harmful chemicals forbidden in many developed nations still occur in China:

Toxic chemicals have caused many environmental emergencies linked to water and air pollution. ... There are even some serious cases of health and social problems like the emergence of *cancer villages* in individual regions [emphasis added].

In a *BBC News Online* report, on February 22, 2013, Wang Canfa, a professor of law and veteran Beijing-based lawyer who defends pollution victims, pointed out that this was the first time the term "cancer villages" had appeared in a ministry document.[2] This first mention of the term in a government-issued document triggered a spate of media reports both in China and abroad (see also Dewey 2013). A search for media articles containing cancer village in the keywords on CNKI (China National Knowledge Infrastructure, the main database of Chinese newspapers and academic publications) in July 2013 identified a total of 207 reports. Among these, only thirteen appeared in 2012, whereas 106 were already published in the first half of 2013, with an additional nineteen by November 10, 2013.[3]

In the ministry document, however, as in most of the media reports, the term cancer villages remained undefined. Given long latency periods and multiple causalities for several cancers, cancer villages are a contested field. Except for cases of acute poisoning and where there is a relatively clear understanding of health effects (as is the case of lead poisoning or for benzene, for instance) establishing causal links between exposure and health effects is complex for several reasons, which, as I outlined in the previous chapter, lie at the interface of science, politics, and socioeconomic contexts. Since the initial appearance of the term, cancer villages have been at the center of a controversy involving the national and international media, a range of government officials and state institutions, non-governmental organizations, natural and social scientists, and the general public, particularly populations directly affected. Most often there is no conclusive scientific evidence to link village-size clusters of high cancer incidence epidemiologically to pollution.[4] If cancer refers to an individual bodily illness, the expression cancer village delineates a shared social life, a space that is at once natural and social. Growing concern and unrest over pollution and its health effects mean that cancer villages are not only a medical phenomenon, but also a deeply social, cultural, and political one.[5]

While cancer villages are by no means the only type of environmental health harm affecting rural China (indeed, disparate types of pollution ranging from heavy metals to fertilizers cause a range of other illnesses), they have attracted increasing attention, and therefore provide a useful comparative canvas against which to understand the three case studies that form the core of the book. In this chapter, I outline a broad typology of cancer villages, including the discourses and actions on the part of local residents, campaigners, polluting firms, local and higher government, the legal system, and the media. To do so, I draw on groundbreaking research by sociology professor Chen Ajiang, China's leading social scientist working on rural pollution and health. With support from China's National Social Science Funding Program and grants from FORHEAD (Forum for Health, Environment, and Development),[6] his research team carried out extensive fieldwork in several cancer villages in the course of over a decade (see appendix).[7]

As with most of my Chinese collaborators and colleagues working on pollution and health, I first met Chen through the FORHEAD network, at the FORHEAD launch conference in 2009. Given the clear overlaps in our research interests and approach, we corresponded closely over the years, and I spent more than a month at Hohai University in 2013 reading work by Chen and his team. They published several articles and a pioneering book titled "Research on China's 'cancer villages'" (Chen et al. 2013), which is currently in the process of being updated and translated into English. Chen generously spent many hours discussing his case studies with me. I am enormously grateful to him for allowing me to summarize his team's findings in this chapter and to elaborate my own analytical stance based on their excellent work. In order to gather additional information on some of the sites studied by Chen and his team, I worked with Chen's former student Dr. Luo Yajuan (whom I also met through FOR-HEAD) to carry out an extensive content analysis of Chinese language media and scholarly articles listed in the CNKI database about China's cancer villages from 2001 to 2013. This also helped to contextualize Chen's cases and to trace the wider development of cancer villages as a politico-semiotic paradigm and investigate the fraught question of evidence (Lora-Wainwright and Chen 2016).

While many of the cancer villages described here present features that very much fit the concept of resigned activism, they also fit the conventional understanding of activism much more easily than the cases I examine in the rest of the book. Indeed, they are much more politically and socially active and visible in their resistance to pollution than my own fieldsites. Although some of the less visible, less collective forms of action are present here too, not having been directly involved in the fieldwork does not put me in a position to comment adequately on these aspects. Instead, I present these cancer villages as a useful context through which to examine the evolution, successes, and pitfalls of rural environmental health activism. On this basis, I extrapolate several elements that influence understanding of pollution and responses to it. These provide an analytical lens to which I return in the book's conclusion, in light of my own case studies.

## The Appearance of Cancer Villages as a Politico-Semiotic Paradigm

Media accounts of cancer villages have had perhaps the biggest, and certainly the most visible, impact on the development of the term. Together with scholarly articles, they describe thirty-six separate cases (several articles of course cover the same case). The first CNKI-indexed article that contained the term "cancer village" was published in 2001 in the Chinese mining newspaper (Zhu 2001) about the village of Shangba (Guangdong). Some of the tropes mentioned in this initial piece are present in many of the future reports on cancer villages, and I found them to be prominent among the phenomena that villagers regard as evidence of pollution in my own fieldwork: changes in the color of the stream, livestock deaths, crop failures, high cancer rates, and men's inability to pass the physical test required to join the army. The article also referred to scientific tests that found high heavy-metal content in the local water and to a list of local cancer deaths.

The number of media reports reached twelve in 2004 and eight in 2005. An article in the state-run *China Daily* on May 10, 2004, headlined "'Cancer Village' in Spotlight" (see also Cody 2004), drew national attention, but officials ordered newspapers to stop covering the story. In 2004, an influential short documentary titled "Rivers and Villages" was aired on CCTV (China Central Television). The documentary suggested that pollution in the Huai river basin caused Huangmengying to become a "cancer village." In subsequent years, report numbers remained relatively stable— between three and eleven per year—with a peak of twenty-eight in 2011 (a third of them covering the case of chromium contamination in Qujing, Yunnan), a fall back to thirteen in 2012, and the soaring peak of 106 in the first half of 2013. This sudden rise is first of all due to the official mention of cancer villages in the Ministry of Environment's document. Soon after this, the state-run news agency *Global Times* posted a "cancer village map" including 247 cases on their microblog (Wertime 2013), accompanied by a weeping emoticon and a caption referring to the Ministry of Environment document. This map, compiled by journalist Deng Fei and based on the cover story "China's 100 Cancer Causing Places" (for *Phoenix Weekly*, April 2009), had been circulating online for years (*Double Leaf* 2009), but this was the first time that it was posted by the state media. There were also additional reasons for the rise in reports. The 12th Five-Year Plan, announced in late 2012, put emphasis on building an "ecological civilization," placing the environment in a position of unprecedented attention, and therefore allowing more coverage of environmental issues. In March 2013, there were also two instances of extremely severe air pollution in Beijing, spurring a rise in media coverage. Finally news items on the contamination of deep wells also heightened concerns about the effects of pollution.

The approaches taken in these reports are predictably diverse. A strong and recurrent theme is the critique of the high importance placed on GDP growth. Articles often contain lists of cancer victims, presented as evidence that cancer is the result of the proliferation of polluting enterprises promoted by the national emphasis on rapid economic development. In most cases, pollution is attributed to official corruption, which poses an obstacle to implementing environmental regulations (*Nanfang Dushi* 2007a, 2007b). This emphasis became particularly prominent in 2008, 2009, and 2010 after: the failure of attempts to measure "green GDP" (taking environmental losses into account when measuring GDP); and the publication of data from a project by the World Bank and China's State Environmental Protection Agency, which showed that cancers associated with water pollution, such as liver and stomach cancer, are well above the world average (2007), though it did not explicitly prove that those cancers were caused by water pollution.

Significantly, however, many reports also promote a more balanced view of cancer villages and stress the remediation efforts underway. This is evident in articles about Huangmengying, in Shenqiu County, Henan province (Deng 2005; Li 2004; as well as the 2004 CCTV documentary previously discussed), a village affected by upstream pollution in the Huai river (China's third largest and most polluted river), which became globally infamous because of its level of pollution. A 2004 news item on Huangmengying highlighted that the government had been making efforts to clean up the area since the 1990s and has now embarked on a new investigation and remediation program. This emphasis on attempts to clean up damaged areas was also present in a piece on Shangba that described ongoing remediation efforts, particularly government investments in a water reservoir and investigation by scientific units as the basis for ecological remediation (Yang and Fang 2005). Later articles on Shangba (Cao 2009; Cao 2013) continued to stress the achievements of remediation efforts and clean water programs funded by the province and city development commission, which provided clean water for more than three thousand villagers. At the time of my survey of media coverage, the most recent piece on Shangba (Cao 2013) highlighted that villagers and local officials alike were keen to shed the "cancer village" label that had brought them so much bad publicity. These developments show a trend for media coverage becoming more mixed than it had been at first, and to highlight government efforts to address the problem rather than simply pointing to its existence and urgency. This suggests that the government, particularly at the higher levels, endeavors to play a growing role in high profile pollution cases. The extent to which it does so beyond these cases is one of the underlying questions of this book.

## A Typology of China's Cancer Villages

Reports over the past decades linked cancer villages first and foremost with chemical industries, and secondly with paper factories, resource extraction and processing, downstream pollution (particularly from industry and mining), and heavy metals (caused by the above forms of pollution). In the majority of cases, pollution sources are multiple and include a range of pollutants, therefore complicating the question of what pollutant may be to blame and which industry may be held responsible. In some cases (such as the Huai river basin), the local geography and geomorphology make it practically impossible to pin responsibility on one specific offender.

Geographically, cancer villages are clustered in coastal regions, especially Jiangsu (particularly southern Jiangsu and the area surrounding Yancheng municipality in northeastern Jiangsu), Zhejiang, Henan (particularly the Huai river basin), and Guangdong provinces, and Tianjin municipality; but some cases are also reported in inland areas such as Sichuan, Yunnan, Hubei, Shanxi, and Guangxi provinces, and Chongqing municipality. The relative concentration of cancer villages in coastal areas—also corroborated by the "cancer village map"—can be interpreted as corresponding to uneven regional development, whereby regions that industrialized earlier display a higher likelihood of pollution. While it is possible, if not likely, that cancer villages (and pollution more broadly) are more common in coastal areas, they are also spreading to historically less developed, poorer inland regions, as these endeavor to catch up with their wealthier neighbors.

In what follows, I briefly describe some of the key case studies analyzed by Chen and his team, in order to identify some commonalities, but also differences, in villagers' strategies, that is, their ability to attract attention and obtain redress. This provides a useful backdrop against which to evaluate the substantive case studies presented later in the book.

## Shangba: Media Storms, Choosing Targets, and the Importance of Cohesive Communities

Shangba is a village of 3,400 residents located in a mountainous area of northern Guangdong province. Villagers have been severely affected by iron mining since it began there decades ago.[8] The province owns the main mine, but villagers themselves have also opened private mines since the start of economic reforms in the 1980s. Due to mining, local water and soil have become contaminated by heavy metals including lead, zinc, cadmium, and copper. Tests on water in the local stream and on the soil revealed heavy-metal content in excess of China's guidelines: lead in the soil exceeded safe standards by forty-four times and cadmium by twelve times. Cadmium is a known human carcinogen, and lead is regarded as a possible human carcinogen, therefore the scientific base for a correlation between pollution and cancer is strong. According to local records consulted by Chen's team, there were 214 cancer deaths from 1986 to 2005, that is, 315 people per year out of each one hundred thousand, more than twice the average cancer incidence in rural China (112.57 per year in one hundred thousand).

Villagers told Chen's team in 2010 that they started to complain about pollution in the 1970s. They initially became aware of pollution's effects because the water was smelly and "poisonous," shrimp and fish died in great numbers, and their skin was irritated if it came into contact with water. Their case attracted the attention of the media: the first report appeared in 2001 (Zhu 2001); with high-profile coverage to follow in subsequent years, these articles in turn attracting scientific and civil society interest to the area (Yang and Fang 2005). Attention by the media and researchers further convinced locals that pollution affected their health and provided them with scientific evidence to support their suspicions. However, as mining was carried out both by the provincial mine and by locals, attributing responsibility remained a contested issue.

The ways in which Shangba villagers tackled pollution was deeply influenced by the social texture of the village. Shangba villagers all shared the same surname, which endowed them with great cohesion and high organizational capacity. The ancestral hall was built near the village committee offices, and, because village cadres were part of the same kinship group, they were trusted, and thus they were effective in coordinating villagers and organizing collective action. This enabled them to organize petitions since the 1980s, to attract media attention, and to file lawsuits. They compiled lists of cancer victims, which were used by the media as evidence of high cancer rates. As a result of citizens' efforts and media coverage, the provincial mine and the provincial and municipal governments offered 14 million *yuan*[9] to build a reservoir to collect rain water and to connect tap water to all homes in Shangba.[10]

The local political economy affected villagers' choice of target and of complaint methods, as well as the responses they received by the local government. The village government supported petitions that requested compensation from the provincial mine, but not from private mines, possibly because of a collusion of interests at the local level. This taught villagers to target their complaints toward the provincial mine rather than those in private ownership. This strategic approach, however, gave the provincial mine director grounds to refuse complete responsibility for pollution, and villagers' requests for compensation and clean irrigation water remained unfulfilled. Shangba's case shows that even when evidence is relatively strong and there are high-profile media reports, results are mixed. If gaining recognition as a cancer village helped to secure clean drinking water, it had negative effects too: it compromised locals' ability to sell their produce in the market and men's ability to find wives. Given such mixed results, during the latest visit by Chen's team, villagers and village officials alike were keen to shed the label "cancer village" and embrace a different identity.

### Dingbang: A Wealthy Cancer Village and Controlled Complaints

Dingbang is a village of 192 residents, part of Xiqiao administrative village, which has 2,639 registered residents and a number of unregistered migrant workers.[11] It is located in historically wealthy northern Zhejiang, downstream from the infamously polluted Lake Tai, some hundred kilometers from Shanghai. The region has supported economic development in the wider Yangtze basin and is at the economic forefront nationally. It has been famous throughout China for handicrafts and silk production for a millennium. Township and Village Enterprises (TVEs) have been the key to economic growth in the past three decades, making many villagers wealthy (most of all through the textile industries). Xiqiao is more industrialized than nearby villages, with ninety-one industries, eighty-one of which are textile industries opened by locals. Per capita income in the 2000s was

12,000 yuan, fairly typical for this region but relatively high for rural China in general. Since the 1980s, TVEs severely polluted local streams and wells, and the environment worsened further with the burgeoning of private enterprises since the mid-1990s. The local pattern of industrialization, with businesses scattered in residential areas, resulted in widespread pollution and made effective environmental monitoring difficult. Conscious of the central role of TVEs and private enterprises in the economic development of the region, the local government turned a blind eye to pollution.

Although villagers admitted that textile industries opened by local families turned the water in the stream white, they argued that the local metal furniture factory was the main culprit for pollution and cancer. This was the largest factory in Xiqiao, its owner was originally from a neighboring village, and it produced garden furniture, mostly for export. The worst pollution resulted from the process of washing, melting, and decorating metals, which relies on sulphuric acid, hydrochloric acid, and oxidization. Until 2005, the industry piped untreated wastewater directly into the local stream. Local residents guessed that the local stream, smelly and black, contained acids and heavy metals, and they blamed it for the dead fish, shrimps, and snails, human skin rashes, and the high incidence of cancer among locals. According to villagers, assisted by the village doctor, from 2000 to 2007 cancer rates (mostly digestive cancers, such as esophageal and stomach) were four times the average for rural China. They regarded the spike in cancer cases two years after the factory's opening, and their decrease shortly after the factory installed a new waste water system, as evidence that the factory was to blame.

As in Shangba, their attempts to gain redress had uneven results. At first they approached the factory, but were not allowed to enter the premises and were unable to contact the manager. Second, they contacted the village and township government, but were told that nothing could be done. Third, they sought help from the former village Party secretary (who served from 1975 to 2001), hoping he would be able to mobilize his knowledge of official channels, protocols, and personal connections to liaise with the factory and township cadres. He had personal reasons to be committed to the cause, too: he had suffered from stomach cancer and was convinced it was caused by pollution. However, he was unsuccessful. Frustrated with the lack of action at the local level, villagers reached one step higher and called the city mayor hotline.

As a result, staff from the district Environmental Protection Bureau visited the metal factory once, but no information on pollution was released. Villagers were told there was no evidence that the factory exceeded pollution levels or that its wastewater could be correlated with cancer. The Center for Disease Control (CDC) is responsible for investigating cases of high incidence of serious illnesses, but this had not happened either. In 2010, CDC staff told Chen's team that in 1992 they began to establish a resource hub for recording illness data, but that abnormally high cancer incidence was not visible at the county level. Because the CDC was unable to assess illness rates at the village level, villagers were trapped between their own village-level evidence and the scale at which the government operates. As a final effort, villagers contacted the local TV station, but the report was never aired, most likely because the local government bribed the TV station. Nevertheless, the factory installed water treatment equipment in 2005, which ensured cleaner water to some extent.

Several features of this case are worth noting. First, it shows that being denied access to data is a serious obstacle to gaining recognition by the government. In Dingbang, as in many other places, villagers were not legally entitled to enter the factory or demand full records of a factory's emissions. Most often, villagers are unable to collect the necessary evidence to support their claims that pollution causes cancer. Second, it shows that local elites, in this case the village doctor and the former Party secretary, are key players in local negotiations. Third, persistence in complaining may have some effects even when villagers' evidence is disputed: although the Environmental Protection Bureau (EPB) and CDC refuted villagers' claims about high cancer rates, the factory was required to install a water treatment plant. Fourth, local political economy and the history of development in the area affect locals' attitude to pollution and their choice of target. The importance of textile manufacturing for local families and its reputation as a key local heritage made it unlikely that it would be blamed for high cancer rates. While rivalries among neighbors may sometimes result in attacks on locally owned businesses, it is more common that opposition would be against those businesses owned by outsiders. Conversely, the metal furniture factory was among the top tax earners for the district, and its boss was powerful and well connected. Therefore, opposition by villagers was not met with support by the township or the city.

Finally, Dingbang's case is a classic example of the uneven distribution of the costs and benefits of industrialization. The biggest benefactor is the metal furniture factory's boss, who earns from his business, but does not live locally, and therefore does not suffer pollution's effects. Factory workers benefit by earning wages, though they suffer pollution by living locally. Local families who run textile factories also benefit from relatively high incomes. These families are more likely in a position to move out of the area when they feel pollution is too severe. At the bottom of the ladder are migrants, unskilled workers, farmers, and the elderly, who have low incomes and are unable to leave the area. Because locals refuse to buy farm produce from Dingbang given its reputation as a cancer village, Dingbang's farmers have no choice but eat it themselves. This concentrates the effects of pollution on those left in Dingbang who still rely on agriculture. Poor families are also unable to dig private wells to access cleaner water, or to buy bottled water, and continue to rely on the polluted village well.

### Jian'nan: A Poor Cancer Village, Petitions, and Mixed Outcomes

Jian'nan is a village of roughly one thousand residents situated in southeast Jiangxi province, five kilometers from the county town.[12] The region is much poorer than Jiangsu, and the village itself is much poorer than Dingbang. It was established as a state livestock farm in 1956 and a land reclamation site in 1959, which over time grew to include a soy sauce factory, a monosodium glutamate factory, a distillery, a caustic soda plant, a fertilizer plant, and a glass factory. Due to poor operation (and the failure of state farms as a whole), most of the factories in Jian'nan closed in the 1990s. Only the glass factory and the fertilizer plant continued to operate after being contracted to private companies. As a consequence of these closures, many workers left and the population dropped, now consisting mostly of elderly and infirm people.

Villagers already complained about pollution before privatization. Because factories were located close to the village, they polluted the local water. By the late 1980s, villagers thought the water tasted "strange" and wondered if it might affect their health. After privatization in the 1990s, air and water pollution only worsened, and villagers filed petitions to the county, city, and provincial governments demanding cleaner industrial production and piped drinking water. Eventually, they attracted the attention of the county government. In 1999 and 2000, village representatives accompanied local government leaders and members of the county CDC on an investigation of three wells. Water tests by the county and the provincial CDC found excess manganese in one well and excess cadmium in another, but these could not be linked to the nearby factories, and, according to the EPB, they were naturally occurring and common in the area. The Environmental Protection Bureau (EPB) also argued that pollution in Jian'nan was not severe and that the glass factory only produced moderate air pollution. Because of locals' frequent complaints, the county CDC organized an epidemiological study in 2007 and 2008. This involved testing drinking water in the village, including water sources up- and downstream, deep wells, and surface water. As previously, they found naturally occurring high levels of manganese and cadmium, but no evidence of carcinogens that could be correlated with local industries.

Despite the CDC and EPB's arguments that cancer rates were not high and that pollution was not serious, villagers treated excessive levels of manganese and cadmium in the water as "iron evidence" of pollution and its deleterious health effects. They blamed the two privatized companies still in operation (the glass factory and the fertilizer plant) for pollution and for high cancer rates. Cancer lists compiled by local petitioners (and verified by Chen's team) counted thirty-two cancer cases between the years 2000 and 2009 among a village of one thousand people. Many victims were in their thirties and forties; the youngest was eighteen. The cancers included mostly liver, stomach, and blood cancers, the last with an incidence ten times the national average.

In response to the villagers' complaints, and possibly due to pressure from media reports in 2005 and 2008, in 2009 the local government contributed some funds and asked each household for 300 yuan to provide tap water. In the same year, the county government released a notice to the glass factory to clean up within six months or it would be fined or closed. However, in 2010 villagers still complained about pollution by the two local industries. During fieldwork by Chen's team, they stated that the glass factory created such thick smoke it would turn their laundry black. They also blamed the "salty" drinking water on pollution. In 2010, fifty villagers locked the glass factory's gate to disrupt production. The head of the former state farm came forward and promised a resolution, but soon after he and the Party secretary went into hiding. Villagers went to the factory again and were offered fifty yuan for each protester who agreed to stop, but they refused.

It is worth reflecting on the identity of villagers who organized petitions and on their choice of target. Two most prominent figures among the five petition leaders were men now in their sixties, Zhou and Xu. Xu was a former state farm cadre and the most educated of the five. His education and previous role as a cadre served him well in organizing villagers' efforts. Zhou, by contrast, was a barely literate retired worker from the state farm. What he lacked in education, however, he made up for in social networks. His son and some relatives held positions of responsibility in government, which the villagers thought rendered him relatively immune to police intimidation and boosted his confidence. Like the former Party secretary in Dingbang, he had deeply personal reasons to join the efforts too. His second wife and her son had died of liver cancer, which he blamed on pollution from the glass factory. His house was also adjacent to the factory walls, and air emissions affected it severely. Their ongoing attack on the two remaining privatized industries is rooted in the local political economy. As most residents were former state farm workers, they felt let down by the failure of the state farm, now having to live on a meager pension of 500 yuan per month. For this reason, locals resented the private businesses that stepped in to fill the vacuum left by the failed businesses and benefit themselves economically with little regard for the environment. The decision to target industries that were still in operation may have been premised on the villagers' disenfranchisement, as well as on the sense that, given the local state farm had failed, pinning cancer on it would be most likely be pointless for the purposes of demanding a cleaner environment or gaining compensation. Blaming pollution on current private bosses may offer more hope.

The question of whether residents' opposition to pollution was successful is a thorny one. Petition leader Zhou was frustrated by the limited improvements, but he still believed that petitions and protests were the only way to ensure a response. Villagers persisted with petitions, despite uneven outcomes, because they remained convinced that pollution was severe and that high rates of cancer were caused by it. Through their efforts, villagers succeeded in gaining the attention of the county CDC and EPB and having water tests and an epidemiological study carried out. But ultimately these investigations were used to dismiss villagers' concerns. Positive changes may in fact be incidental side effects of state policies rather than direct outcomes of pressure by the local community. The provision of piped drinking water in 2009 was part of local rural development initiatives, including building a new public road. Similarly, the EPB's acknowledgment that the glass factory should not be situated in a residential area and promises that it will eventually move to the county industrial park are due to new zoning laws and integrated rural-urban planning regulations.

### Likeng: The Power of the Media and Compromised Evidence

Likeng is the name of a small water reservoir selected in 1989 for a waste burial site, and more recently the site of a waste incinerator.[13] The surrounding area became known as Likeng Village through media reports about local pollution in 2009. Likeng is part of a relatively large, industrialized, and wealthy village in Guangdong province. Most of the village land is rented out to factories, and the remaining land is contracted to migrants for farming. The main sources of income for locals are land rental fees and room rentals to migrants. Most young adult villagers work outside the area, leaving behind only the elderly and children. Pollution began to affect the area when the waste burial site was built. The capacity of the wastewater treatment was insufficient, and therefore water in the nearby reservoir was black and smelly, and the fish died. During heavy rain, wastewater flowed into the road, fields, local streams, and fish ponds. The burial site was sealed in 2005, and an incinerator was established near it. Villagers protested and blocked the road, but the plans went ahead nonetheless. During interviews with Chen's team in 2011, they complained that the incinerator emitted black smoke and left a layer of black, poisonous dust containing dioxins on the nearby homes. They argued that the smell was worse at night, when plastic and other chemical products were incinerated. They also complained about wastewater contaminating their fields. They stated that as a consequence they did not dare eat locally grown crops, and they rented farmland to outsiders who sell the crops elsewhere.

Contention escalated when plans were made to build an additional incinerator within ten kilometers of the nearby gated communities in the Panyu district of Guangzhou. Panyu residents were generally wealthy and there were several journalists living in the area. In October 2009, they posted a report online including a list of recent cancer deaths they attributed to the existing Likeng incinerator. According to the list, there were forty-two cancer cases between 2006 and 2009 (fifteen times as many as before the incinerator was built), thirty-four of them respiratory tract and lung cancers, which allegedly proved they were a result of air pollution. The following month, roughly one thousand Panyu residents went for a "stroll" (a code word for protest) to demand that the city government stop the plans to expand the plant. Days later a report was aired on state-run CCTV about incineration and cancer, and an article was published in *China News Week* titled "Likeng, We Are Sorry for You."

The media storm had several effects. Local officials declared in December 2009 that plans had been suspended, that a new environmental impact assessment would be carried out, and more public consultation would take place. Eventually, in 2012 the government announced the project would be moved to a new site roughly twenty kilometers from Panyu, and construction began in 2013 (Johnson, 2013a). The existing plant was required to increase transparency and allow citizen supervisory groups chosen by villagers to enter the plant at any time and gain access to emission records. In a sense, Panyu residents' protests succeeded in attracting attention to Likeng, although it remains unclear whether pollution decreased. For their part, Likeng villagers did not trust that any data they could access would be reliable. They also did not believe that the plant operated incineration technology properly and thought that plant managers would cut corners and lower incineration temperatures to save money. They felt pollution was still severe. Government action then did little to improve villagers' trust in the government, the incineration plant, and their views of pollution.

Local cancer rates are a particularly controversial point in Likeng's case. Following the media storm, the local government's Center for Disease Control investigated cancer incidence surrounding the Likeng incinerator. The CDC found that the media report omitted twenty-seven cancer cases between 1993 and 2005 (before the incinerator was built), and that once those were included, there was no statistically relevant difference between cancer rates before and after the incinerator was built, and no significant increase in respiratory tract cancers. There were other inaccuracies too. Chen's team crosschecked the cancer name list circulated by the media with the families of the victims. Significantly, they found that many had never been consulted by journalists. In one case, the age of the cancer victim was wrong. Twelve of the records that appeared as respiratory tract or lung cancer on the cancer name list were actually liver cancer, stomach cancer, and brain tumors. While we cannot know for certain why these inaccuracies appear, it is likely they would have been introduced by journalists to lend weight to their allegations that the incinerator caused high cancer rates.

This case makes evident the crucial role of media pressure in successful resistance. The media began to focus on villagers' concerns about local pollution and cancer rates only once residents of the wealthier and more media-savvy Panyu district became concerned that incineration would affect them. This case also highlights the comparative success of wealthier Panyu middle classes in opposing pollution, while villagers living in the immediate vicinity of the plant continued to feel hopeless about decreasing the levels of existing pollution. The fact that the additional incinerator was eventually built in a poorer rural area further demonstrates the more severe effects of pollution on those who lack the networks, power, or financial capital to oppose it successfully or to escape it (see Johnson 2013a).

### Dongjing: From Violent Protests to Petitioning and Lawsuits

Dongjing is a village with more than two thousand inhabitants in northern Jiangsu, a region that is historically underdeveloped as compared to the surrounding areas (such as southern Jiangsu, where Dingbang is located).[14] Struck by an inferiority complex of sorts—what Chen (2010) described as "secondary anxiety"—and eager to catch up economically with the rest of the province, the local government is keen to attract investors and prepared to cut environmental corners to do so. In 2000, the township government that presides over Dongjing made a deal with a chemical industry, Julong, which started to operate in Dongjing later that year. Local residents told Chen's former student Luo Yajuan during her fieldwork in 2007, 2008, 2009, and 2011 that they had agreed to the establishment of a chemical industry in Dongjing because they had been told it would not cause pollution, but they soon realized that the industry had lied. They recalled that the air became smelly, thick, and unbearable, fish and shrimps died, the water tasted strange and turned red. They did not dare eat their own food because it tasted of chemicals. Julong's main products were phloroglucinol and o-chlorophenol. Both are organic compounds and derivatives of phenol, the first used in pharmaceutical production and the second in pesticide production, and neither is clearly correlated with increased rates of cancers. However, villagers held local pollution responsible for local cancers. Water tests that found chloride in excess of safe standards by two thousand times provided them with what they considered "iron evidence" of serious pollution.

Angered by the factory's behavior, in 2001 villagers surrounded it to stop production. As Julong ignored them, some began to damage the industry's chimney, windows, and wastewater pipes. The police soon arrived and arrested several people, and threatened that they would not release them unless they promised not to protest again. This experience taught villagers that direct confrontation tactics were not successful and that the township government was on the side of industry, not of villagers. Instead, they began to collect evidence of harm and pursued peaceful methods to complain. A key figure in this was Duan, a villager in his sixties with a rich life experience and some understanding of the law. Duan had not joined the protests. He made meticulous calculations about the losses incurred by villagers and collected records of cancer deaths between 2001 and 2005, according to which eighty-three villagers (out of 2115) developed cancer, sixty-four of whom died. This is more than five times higher than the average for rural areas. Most cases were digestive cancers (fifty-four), the next highest being esophageal cancer (forty). Between 2002 and 2004, Duan, ten village party members, and twenty villagers phoned the county EPB to report pollution and made countless petitions to the county government and to provincial levels, but received no redress. This alerted villagers that EPB staff at the county and city level also denied responsibility, and they would inform the township before they carried out investigations, allowing the industry to temporarily comply with regulations during the EPB visit. Duan concluded that: "officials protect each other."

Following a water contamination accident in March 2005, villagers led by Duan tried to contact the local media, but the program was never aired because the township bribed journalists. Having lost faith in the local government and the media, Duan petitioned the central government. This attracted a journalist from Beijing who did not accept any bribes despite vigorous attempts, and who published an incriminating report in November 2005. Many journalists visited the area soon after, but eventually the media interest died down, and villagers felt that the media too had failed to solve the problem. No government bureau intervened to control pollution. Frustrated by the lack of action, villagers decided to take the county EPB to court for not fulfilling its responsibility of controlling and investigating the level of firms' pollution as stipulated by law. A "barefoot lawyer" from the neighboring town agreed to be their attorney.[15] In March 2006, Duan and three others filed an administrative lawsuit demanding that the county EPB should monitor Julong and ensure they abide by environmental protection. The county government responded by requiring Julong to stop production by May 31, 2006 and to move to the county ecological chemical park by the end of 2007. Villagers, however, were still not satisfied. Although Julong moved, it never compensated villagers for all the damages it had caused in previous years. Despite intimidation and offers of bribes by the town and county government, Duan and three others acted as representatives (attorneys) of 369 villagers and filed a civil lawsuit at the city people's court, demanding that Julong pay losses of 738,032 yuan (about $118,000 at the time). The sum was based on detailed calculations about spiritual damages to relatives of cancer victims, compensation for lost vegetables, and for rice. However, the court required the villagers to present evidence of these losses and proof that they were caused by Julong, and on these grounds the villagers lost their case. Their appeal to the provincial court was met with the same outcome.[16]

In Dongjing, as in several of the cases described earlier, results were mixed: the polluting company moved, but villagers failed to gain compensation for past losses. The importance of the local political economy and of local protectionism is transparent. Local industry contributes a large share of local government revenue through taxation and through fines for pollution, therefore the local government is unlikely to regulate it too stringently. The structure of government responsibility also contributes to the lack of redress: when higher levels of government are alerted to local misconduct, they simply contact lower levels and tell them to sort it out. In other words, whether problems are solved ultimately depends on the local government.

Dongjing is distinctive as compared to villages like Shangba or Huangmengying (see below), because there was a single, clear polluter. This should make the question of responsibility for pollution straightforward. However, villagers were unable to prove Julong liable to pay compensation. This prompts some reflections about the role of litigation. While administrative litigation against the local EPB resulted in relocating Julong, civil litigation against Julong in pursuit of compensation failed. Villagers were unable to turn pollution's effects, which for them were a social fact, into a legal fact.

In such circumstances, the law ultimately makes their suffering invisible, caught as they are in a double bind of being victims of pollution but unable to achieve recognition of their suffering (see Phillips 2012). This begs the questions: If courts do not take into consideration the ability of villagers to compile the kind of data required, who or what are they protecting? Are they actually safeguarding justice? (Luo 2013). Although environmental litigation in China has grown in recent years (Stern 2013), it is little wonder that lawsuits are not villagers' first resort, and that they employ many more diverse strategies. Dongjing's case suggests that through persistence and laborious collection of evidence under the leadership of a skillful villager, villagers may be able to obtain some redress, but that obtaining compensation is a thorny and elusive business.

### Huangmengying and the Huai River Basin: Strong Civil Society and High-Level Attention

The most striking example of the power of the media and civil society to affect cancer village cases is Huangmengying and the wider Huai River Basin.[17] The Huai River is naturally prone to flooding, and has been the object of several hydraulic projects since the 1950s. However, as industry boomed in the area with the advent of economic reforms in the 1980s, this system of reservoirs and canals aimed to prevent flooding and allow irrigation also spread polluted water in the surrounding area. For this reason, it is also difficult, if not impossible, to trace pollution to specific sources and industries. An accident in 1994 attracted the attention of the central government that invested in a multibillion yuan pollution control project. However, similar accidents took place in 1995, 1999, and 2004, pointing to the inadequacies of the project and to continuing pollution.

Long before the State Council (China's cabinet) took notice, villagers already suspected a connection between cancer and water pollution, and they were drilling wells as early as the 1980s to avoid having to drink from the river. Newspaper reporter and photographer Huo Daishan played a key role in their efforts to obtain redress. A resident of Shenqiu County in Henan, home to China's largest cluster of cancer villages on the Huai River, Huo set out to document local pollution in 1987, soon after local residents began to notice it. Huo turned to campaigning full-time in 1998 and formed a group called "Guardians of the Huai River" in 2000. These efforts attracted high-profile media attention, and led to major central government investigations and remediation projects. The report "Rivers and Villages," aired on CCTV in 2004, suggested that water pollution in the Huai Basin had severe effects on public health, causing a spike in cancer rates.

Civil society has played an important role in cleaning up the region. Huo and his group have trained hundreds of volunteers who now work in teams to regularly monitor the river and conduct water-testing, pushing companies to implement pollution-control measures. They have mounted dozens of exhibitions in cities, universities, and villages; they have written letters exposing the illegal activities of local officials and factory owners; and they have raised awareness of local pollution through the media and conferences. They have installed deep-water wells and low-cost water filtration systems in local communities, with support from government and private institutions, and provided hundreds of cancer patients with medicines. They have also built cooperative relations with local authorities and industries, most prominently a local monosodium glutamate (MSG) factory, one of China's largest MSG manufacturers and a major polluter. In recognition of these achievements, Huo won the prestigious Ramon Magsaysay Award in 2010.

Local and central governments have taken actions to tackle pollution in the Huai river basin in a range of different ways, doubtlessly spurred not only by the gravity of the situation, but also by the increasing pressure created by civil society. After the report on China's central television in 2004, the central government designed nation-wide and locally based programs for cancer prevention and treatment, and it tasked the national Center for Disease Control and Prevention with an investigation on cancer and water pollution in the Huai Basin that would last for eight years. The findings of their study were published in the *Atlas of the Water Environment and Digestive Cancer Mortality in the Huai River Basin* (Yang and Zhuang 2013): a collection of 108 maps showing the severity of water pollution in the Huai River and its tributaries, and rates of digestive cancer deaths in the region. For the first time, they officially correlated high rates of cancer to pollution in the Huai River Basin. Because the scale of the study is sufficiently large and aggregates data at the county rather than village level, it has statistical validity. Polluted water, the study suggests, may be directly blamed for several especially prevalent cancers of the gastrointestinal system in the area, including those that attack the esophagus, colon, and rectum.

Some campaigners in China have expressed hope that the newly published atlas could be used as evidence in future environmental public interest lawsuits aimed at winning compensation for victims of pollution (Wang 2013a, 2013b). The atlas, however, does not provide evidence to link specific pollutants or industries to health effects in particular villages. Indeed, it is typically the lack of this information that hinders villagers' and campaigners' efforts to obtain redress from polluters. The fact that pollution has multiple sources and that local flood control and irrigation systems spread it across wide areas further complicates attempts to pin down particular polluting firms for damage in specific sites. Nevertheless, the atlas has created the institutional momentum to close polluting factories in the area and to support further clean up initiatives, even if it does not offer the evidence that would be required to deem specific polluters responsible for health effects and require that they compensate local populations.

This outcome suggests that high-profile attention is crucial in obtaining intervention, even when specific causal links between pollution and health may be unproven. By contrast, villagers who demand redress in areas that received less attention by the central government and the media are much less likely to obtain intervention. And yet, even in areas like Huangmengying, which have been the focus of so much media and central government attention, results are mixed. Media reports and central government initiatives spurred action at the local level. During their fieldwork in Huangmengying, Chen's team found that the government built three deep wells in the village to supply clean drinking water to residents. These wells served an area with 40,000 people, including those outside the village. Shenqiu County authorities took action in 2005 by connecting village water supplies to forty-seven newly dug wells. Twenty villages with the highest cancer rates were among those benefiting. This was part of a wider initiative by the State Council to provide clean water to rural areas as stipulated in both the 10th and 11th Five-Year Plans (2001–2005 and 2006–2010). Typically this initiative would not have been applied to villages like Huangmengying because, according to regulations, pollution of drinking water should be resolved by the parties responsible for causing pollution. That it was applied successfully here is due to the notoriety of the site, which put pressure on the local government to solve the problem and secured higher government funding to do so, in the absence of evidence of liability for particular firms.

All these efforts have improved the situation, but the Huai River is still far from clean (personal communication to Huo Daishan, November 12, 2013; Wang 2013b).

## Analysis and Conclusion

The six cases described above suggest that residents of cancer villages respond to pollution in a variety of ways, including protests, petitions, lawsuits, approaching the media, NGOs, and a range of experts, lobbying polluters and various levels of government, collecting their own evidence, demanding compensation, or resigning themselves to the situation and taking individual or family-level measures to minimize its effects. Villagers often first try to liaise directly with polluting firms; second, they call on various levels of the government and organize petitions (starting from the local authorities and climbing the administrative ladder if local offices are not responsive); third, they contact the media, NGOs, and a range of 'experts'; and, finally, if they are still dissatisfied with the outcome, they may resort to lawsuits. However, villagers' actions do not always follow this timeline: some strategies may be visible at different times in different localities, or they may coexist in the same site.

Research on cancer villages highlights that several factors affect the ways in which individuals and local communities perceive pollution, and consequently the timing and types of action embraced, their outcomes, and the responses such actions trigger on the part of local and higher levels of government. These factors include the following:

(1) Types and levels of pollution, relative clarity of its link with particular illnesses, level of awareness of pollution and its harm.
(2) Community cohesion, its organizational potential, and the role of charismatic leaders (particularly local doctors and elites, including village cadres, and villagers who are well-connected, educated, or have had a rich life experience outside the immediate area).
(3) Local political economy, degree of dependence on industry, and relationships to various levels of government.
(4) Support from civil society, the media, and outside expertise.

While the relevance of these various factors may vary in each case, we may draw several general conclusions. First, the demonstrability of a link between pollution and cancer seems to have a relatively limited impact upon which places gain attention, recognition, and redress. One important reason for this is that not all cancer villages display an equally strong relationship between pollution and cancer. Chen and his team are commendable for their efforts in establishing the uneven extents to which evidence of a link between cancer and pollution existed in each of their sites. This required integrating attention to environmental data with a study of individual and social factors that may also contribute to the onset of cancer and other diseases (Chen 2013; Holdaway and Wang 2013). In addition, evidence of the correlation between cancer and pollution, even where it might be scientifically backed, is extremely hard to obtain. Shangba is the only one of the six cases I examined where the correlation between cancer and specific pollutants (such as cadmium) derived from local firms is relatively clear. In the Huai River Basin, the correlation is also strong, but cannot easily be traced to particular firms or to the village level. In the four remaining cases, villagers' claims that pollution causes cancer were dismissed by EPB and CDC studies. In part, this is because the scale at which villagers collect data is too small and the time period too short to be recognized by EPB and CDC as statistically significant. Because of this gap between the micro-scale of villagers' experience and the level at which official data is aggregated, villagers' data is rendered irrelevant.

For villagers, however, water tests that showed certain elements to be in excess (even if they were not proven carcinogens) were unquestionable proof that pollution is severe and that it is linked to higher cancer rates. Despite bureaucratic and scientific dismissal of what villagers may regard as painfully obvious, their persistent complaints become evidence of harm, or at least threatening enough to social stability that the government opts to intervene. In this sense, cancer villages remain largely a socio-political phenomenon rather than an established medical fact. The term cancer village and the lists compiled to support the term operate as "weapons of the weak" (Scott 1985), politically sensitive practices that serve to attract attention and remediation where the local community has otherwise little or no power to do so. Indeed, residents of the four villages where evidence of a correlation between local pollution and cancer rates is lacking still received some redress, such as the installation of water treatment plants (Dingbang), piped drinking water (Jian'nan), cancellation of plans for the incinerator in Panyu (though the one in Likeng remained in operation), and the relocation of the polluting firm (Dongjing). While the village is a "very difficult unit of analysis" for establishing epidemiological correlations, it is undoubtedly an important unit of social life, and therefore is likely to remain a unit of focus and contestation for rural communities (Holdaway and Wang 2013).

This highlights the second point: the importance of villagers' ability to raise concerns over pollution and particularly their persistence in the face of polluters' and local governments' refusals. In Shangba, kinship was a vital organizational force, and, in all other cases, key local figures emerged who drove forward efforts to seek redress (though in Likeng's case, these were residents of Panyu rather than the village in which the incinerator was located). In the Huai river basin, Huo Daishan (an educated and well-connected local journalist/activist) played a crucial role in attracting attention and funding from the central government and putting pressure on local polluters to clean up. In several of the cases studied by Chen's team, village doctors acted as resource hubs for the community (and for researchers) and helped to gather evidence and compile cancer lists. Finally, supportive cadres (as in Shangba) or former cadres (as in Dingbang) helped to sustain villagers' demands. In sum, the presence of charismatic leaders and of a cohesive community played a key role in affecting the types of action embraced and their outcomes. Environmental protection is seldom enforced strictly in areas with polluting industries unless the local population demands it. Without such pressure, businesses resume operations as soon as the pressure shifts. In several cases reported in the media, government action is most commonly to fine the industry and secure some compensation for villagers rather than to clean up or stop pollution entirely.

Third, local political economies affect the focus of villagers' complaints. In Shangba, villagers targeted the former state-owned mine rather than private mines run by locals. In Xiqiao, they targeted an industry owned by an outsider rather than the numerous textile factories opened by villagers. In Jian'nan, villagers focused on the two privatized firms rather than on the lingering pollution caused by the former state farm. In all these cases, pollution caused by villagers themselves was overlooked in favor of focusing on external parties. Local political economies also influence the response to villagers' demands. In Dongjing, Julong's importance as a source of tax revenue protected it from tighter regulations until the EPB was sued for malpractice. When pollution is exposed to higher government levels, the common outcome is that pressure is placed back upon local governments to maintain social stability and deal with pollution. Given the financial pressures on township and county governments, they may opt for limited pollution monitoring and repression of local protests. In a few cases—where a media storm successfully attracts too much attention to the region for the status quo to continue undisturbed—pressure and funding from higher levels enable and require township and counties to monitor pollution and to engage in remediation initiatives, as was the case in Shangba and Huangmengying. Even in such cases, however, the results are not entirely satisfactory for villagers, and redress may only be partial.

Finally, outside support, particularly from the media, seems to improve the chances of government action, although its effects are also uneven. All villages, except for Jian'nan, were the subjects of varying degrees of media attention. National media coverage of Shangba and Huangmengying no doubt aided villagers' efforts to gain attention and redress. Local media outlets, however, are less likely to play an important role. Indeed, in Dingbang's case, the local TV station never aired the report. Dongjing also faced a similar setback, and only journalists from other regions reported on the case. Conversely, the media has the potential to create a controversy and succeed in halting plans for potentially polluting plants even when the evidence they cite is in fact inaccurate, as in Likeng's case. However, the power of the media is temporary in most cases. The pressure the media creates lasts only as long as a case is in the spotlight. Whether promises made by officials and firms when their locality is under scrutiny actually materialize and last in the longer term is more likely to depend on the previous two factors: the persistence of the community in complaining and the local political economy.

Of course, villages are not homogenous entities: pollution has uneven effects on different social groups and their responses vary accordingly. Often, industrialization results in more pronounced socioeconomic stratification, and, conversely, differences in social and economic capital affect the extent to which individuals and families are exposed to pollution and in how they react to it. Those who can afford it, try to minimize their exposure to pollution by digging private wells, buying bottled water, or moving to cleaner areas. In severe cases, only the poorest, elderly, and infirm are left behind. This highlights the inherent inequality and injustice in how pollution affects local residents: wealthier residents (who often became wealthy through polluting businesses) are later able to escape pollution or to protect themselves from it, as was the case for the factory owner in Xiqiao. Poorer residents by contrast are left with little choice but to endure it.

Given that my analysis here is based on a review of work by Chen's team rather than on my own fieldwork, I am not in a position to elaborate on the subtleties of locals' subjectivities or to present these sites as cases of resigned activism. Rather, they provide a useful glimpse into the obstacles to rural environmental health activism, even where more contentious forms of action are embraced and where the sites gain relative visibility through media coverage. Given that the success of activism remains mixed at best, even in more visible and more politically active sites, more attention should be granted to localized, less collective, and less visible engagements with pollution. On this basis, the cases examined in the rest of this book high-light when, how, and why pollution becomes normalized and with what consequences.

# 3 "Toxic Culture": The Spectrum and Origins of Resigned Activism

## Introduction

"Don't think about things you cannot change."
—Guo Lin, thirty-one-year-old man with a low income, informal worker at Linchang, May 16, 2009.

"We're old enough. So we are not afraid of illness. The worst that can happen is death, so what?"
—Mr. Zhang, migrant from Huize in his forties, informal worker at Linchang, May 6, 2009.

"If you run around complaining to the government about pollution, they will say you have [mental] problems ... so people don't go. The only way [to avoid pollution] is to move out."
—Fifty-seven-year-old former teacher and wealthy owner of a local factory, July 9, 2009.

On May 2, 2009, I boarded a coach in Yunnan's provincial capital, Kunming, bound for the Linchang factory in Baocun village.[1] The writing on the side of the coach ("chemical factory") suggested this service was most likely provided, at least initially, for the factory's formal workers, many of whom lived in Kunming during the weekend. I observed the other passengers in anticipation. Setting up fieldwork in Baocun had taken months of negotiations enabled through contacts provided by Benjamin Van Rooij, a scholar of Chinese law and project collaborator who had carried out research in Baocun some years previously (Van Rooij 2006).[2] I was extremely eager to finally be able to settle there with three Chinese researchers. Fellow travelers struck me as well dressed and quite young, but probably not all that healthy. Many of them carried plastic bags literally full of medicines, including painkillers, tonics for the liver and kidneys, and various remedies for rheumatism and arthritis. Zhou, the public health specialist on our team, informed me that these ailments are closely correlated with exposure to phosphorous. As the coach approached the township and industrial complexes began to grow in numbers, the air became increasingly pungent. As we drew nearer and nearer to the final stop in the village, the dust from nearby phosphorous mines and emissions from local processing plants was mixed with the thick dust created by the heavy traffic on the paved but sandy road. The acidic cocktail produced by this industrial environment stung my throat and nose. I tried to ignore it as I walked with my hosts toward their home just in time for dinner. The village was crowded with buildings, every scrap of land occupied either by locals' homes or by simple, small concrete dwellings rented out to the large migrant population that gravitated toward Baocun attracted by work opportunities. The communal toilet and garbage dump located in the midst of village housing exuded an overpowering smell. Scarcely able to make it past the entrance, during my stay I often resorted to a disused outdoor pit adjacent to a pigsty (the common setup in rural homes), long since abandoned as the local economy shifted toward industry. The village itself seemed much dirtier, dustier, and noisier than the more "traditional" agricultural village where I had lived in Sichuan province in 2004 and 2005. I had never seen quite so many flies. As the excitement waned, I became aware of a dreadful headache. When I blew my nose, I noticed some specks of blood. In the coming days, I learned from local residents that this is "normal" in Baocun, especially for new arrivals. Zhou confirmed it was most likely a reaction to the high acid content in the air. Later that evening I wrote in my fieldnotes "this looks like hell on earth." This chapter describes how those who live in Baocun experience and make sense of these conditions, and how they have gradually accepted them as part of life. It also reflects on how the burden of environmental health harm is unequally distributed among the local population, shedding light on some of the injustices intrinsic to the capitalist path toward industrial development.

## Baocun's "Toxic Culture": Socioeconomic Stratification and Its Effects

Baocun is a heavily industrialized administrative village in Yunnan province (southwest China), reliant on the mining and processing of phosphorous. In 2009, the total registered population was over two thousand, but the village also accommodated ten thousand migrant workers attracted by work opportunities in the local industries and mines. It is divided into five subvillages (roughly mapped onto natural villages), with mixed surnames. The main industry in Baocun is Linchang, a very large, formerly state-owned fertilizer plant that opened in 1982. In 2001, Linchang was transformed from a state-owned enterprise into a joint-stock company, still partly owned and administered by the provincial government, but also, from 2005, forming part of a multinational corporation. It produces phosphoric acid ($H_3PO_4$), sulphuric acid ($H_2SO_4$), ammonium dihydrogen phosphate ($NH_4H_2PO_4$), potassium sulfate ($K_2SO_4$), and sodium hexafluorosilicate ($Na_2SiF_6$). Alongside Linchang, there are a plethora of privately run mines and more than a dozen small and medium industries that provide much of the supporting infrastructure for Linchang's resource extraction and processing (see figure 3.1). Altogether, industries contribute seven million yuan per year in taxes to the state, and one per cent of this is given to the village.[3] Industries also pay land rental fees and pollution fees, and provide employment opportunities. Linchang provided much of Baocun's infrastructure: roads, a train line (for industry products), a good middle school, the best medical care in the area, and free irrigation water and drinking water (though it does not reach some areas of the village). Therefore, over the past thirty years, Baocun's social, political, and economic life has become inseparable from local industries.[4]

Intense and growing industrialization in Baocun has brought about increasing stratification and inequalities. Benefits and costs of industrialization are divided along several overlapping axes: spatial, social, economic, and political. In spatial terms, subvillages and the formal workers' living quarter were diversely affected by pollution and unevenly entitled to compensation for accidents, yearly pollution fees, and land rental fees. Residents of the living quarter were not entitled to any of these fees, but they were also less affected by pollution: the wind usually blew from the industrial area away from the living quarter, and their drinking water was deemed safe by tests, unlike many other water sources in Baocun. The subvillage of Tacun lost most of its land to Linchang, and therefore received the highest yearly compensation, totaling 150,000 yuan per year in 2010 and due to rise every three years. Sancun by contrast only received 20,000 yuan per year and Qingcun only 5,000. Pollution fees also varied depending on the quality of the land affected, with the highest amounts given to

**Figure 3.1**
Bird's-eye view of some of Baocun's local industries, 2009.

Tacun residents. Other firms located in each subvillage offered small gifts to elderly residents in their respective subvillages, usually between one hundred and two hundred yuan per person per year. As a consequence, residents of subvillages with more industries and mines received higher amounts. Finally, subvillages contracted uneven quantities of land to mines, and therefore residents received different sums in land rental fees. Once again, Tacun residents received the highest amounts, peaking at 15,000 yuan per person in 2008 and 8,000 in 2009. These are very high payouts, even for residents of an industrialized village, considering that informal workers might earn as little as thirty yuan per day. If Tacun drew the highest economic benefits, it also suffered the worst effects of pollution. What little land remained was by and large too polluted for agriculture. By contrast, subvillages further removed from industrial premises, like Shaocun, still relied heavily on farming. Overall, industrialization deepened the differences between these microregions and their residents.

In political economic terms, Baocun is at something of a crossover between China's southern "sunbelt," which relies overwhelmingly on migrant workers, and its northeastern "rustbelt," where the workers of former SOEs (state-owned enterprises) predominate (Lee 2007). Baocun's Linchang is also a former SOE, but the first generation of formal employees was composed of villagers whose land was expropriated and who are now in receipt of a good pension. They are therefore unlike the disenfranchised laid-off workers in the rustbelt (see Hurst 2009) and feel relatively loyal toward Linchang. On the other hand, like employees of firms in the sunbelt, the vast majority of Linchang's workforce is made of poorly paid, low-skilled workers. While many of these are migrants, roughly 50 percent of locals are also employed in Linchang and other local industries. Indeed, when the first generation of formal employees retired, they were replaced by more educated outsiders rather than by villagers. Younger villagers (who sought employment after the 1980s) lamented their inability to secure formal employment as their parents had done, but still benefitted from a range of income opportunities locally, including informal employment.

Predictably, socioeconomic groups are diversely positioned to benefit from industrialization and to suffer the effects of pollution. In turn, industrialization has deepened these socioeconomic gaps over time. At the top of the proverbial pyramid stand local elites. These include those with official positions in the village and subvillage governments, and owners of local firms, such as industries and mines (see figure 3.2). Often, political and economic capital was concentrated in the same hands. For instance, village leaders have negotiated positions of power through deals with industry bosses and by managing the Labor Company, which arranges all contractual work at Linchang. In the subvillage of Sancun, one strong kinship group has historically held most official positions and much economic power. Three brothers by the name of Wang owned some of the village firms. One of the brothers, while serving as subvillage head, embezzled 750,000 yuan that had been given to the village by Linchang, and was still able to escape conviction through some powerful connections. In his case and in some others, there were rumors that villagers were bribed for their votes. Wealthy villagers, like the Wang brothers, no longer lived in Baocun, having moved in search of a more comfortable (and healthier) home.

**Figure 3.2**
Home of a member of the local elite, 2009.

The next layer of the pyramid is made of skilled formal workers who reside in the salubrious living quarters (see figure 3.3) and do relatively safe jobs. They are paid between 2,000 and 4,000 yuan per month, but also receive large bonuses. An electrotechnologist, for instance, might earn 50,000 yuan per year. Formal workers also have workplace insurance, and they are offered a yearly health checkup. One step down, we find registered residents of Baocun. As I explained, they are a very diverse group in terms of spatial divisions, but they are also differentiated in several other ways: at the wealthier end of the spectrum, villagers may run small businesses such as restaurants, shops, and majiang bars. They may also rent rooms and land to migrants. Some older villagers are retired Linchang workers, some of whom also reside in the living quarter and receive a monthly pension of 1,000 yuan. Villagers with good connections may secure lighter work in the local firms through deals with local officials. Less connected villagers are left with heavier or poorly paid work. At the bottom of the pyramid, indisputably, lies the vast migrant population: they are most commonly employed informally, do the large share of

**Figure 3.3**
Living quarter for skilled workers, 2009.

arduous and dangerous jobs, and may earn between 700 yuan and 5000 yuan per month, depending on the type and amount of work undertaken. As becomes apparent, social, economic, and political diversities are inextricably linked, and they intersect with uneven levels of exposure to pollution. The wealthiest have long since moved away, thereby avoiding pollution. Villagers who are relatively well-connected and financially advantaged manage to avoid the most polluting jobs and may purchase bottled water. By contrast, the poorest residents drink local water sources, have no workplace insurance, are not entitled to a yearly health checkup, and take on the most harmful work.

The gradual process that led to the development of such inequalities may be described as a "toxic culture" (Hofrichter 2000). In the context of environmental racism in the US, Hofrichter adopts this term to highlight that toxicity is not simply about pollution but also about "the social arrangements that encourage and excuse the deterioration of the environment and human health" (1). Similarly, in Baocun, lax enforcement of environmental protection and coalitions of interest among the economically and politically powerful have allowed industrialization to entrench inequalities and to expose those at the bottom of the socioeconomic ladder to the most adverse hazards. Baocun migrants in particular are subjected to a toxic culture because as a group they are more exposed to pollution as a result of living closer to Linchang, by doing more strenuous jobs, and yet not being entitled to compensation or pollution fees. But toxicity also results from their uneven structural positions, and the deep-seated and internalized lack of entitlement to live and work in a less toxic environment that is so pervasive among migrants and poor locals. Toxicity, in other words, is not only an ecological phenomenon, but also a social, political, and economic one, involving inferior access to jobs, housing, schools, welfare, and other resources. In this chapter, I map out the different degrees and origins of resignation among migrants and poor locals.

### The Plight of Baocun's Migrant Workers

The social and economic reforms implemented after Mao's death shattered the "iron rice bowl" of lifelong job security and encouraged market competition. Rural-to-urban migration was an essential ingredient to the success of reforms and indeed to the economic rise of China. It tackled the rural labor surplus and readily provided a huge and cheap labor force to the growing industrial sector. More than two hundred millions rural migrant workers (*nongmingong*) have formed a new working class with limited rights and welfare (Chan 2001; Lee 2014; Murphy 2009; Pun 2005; Solinger 1999; Yan 2008; Zhang 2001). Their work is precarious, and suffers from being low pay, low skill, high intensity, often monotonous and repetitive, in poor conditions, and lacking employment protection (Zhou 2013). The experiences of Baocun migrants reflect these well-known facts, but they also articulate a particular kind of hardship: exposure to severe pollution. In what follows, I argue that Baocun migrants are not only in a disadvantaged and insalubrious position, but they have also come to accept their plight as inevitable.

Numbering five to one as compared to the local population, more than ten thousand migrant workers made up the majority of Linchang's labor force and served as the engine of local development. Informal employment in the local industries was administered by the Labor Company, a local government office that in turn supervised a plethora of subcontractors, many of whom were from other parts of Yunnan. Before workers' salaries reached them, they went through both the Labor Company and their subcontractor, and each level withheld a percentage of their pay. Originally, the Labor Company was established to ensure that villagers would have access to industrial work. However, since the late 1980s, as demand for labor exceeded local supply, and, more recently, as some villagers began to look for work elsewhere, the Labor Company also resorted to subcontractors to attract migrants.

Good connections were instrumental in securing work at Linchang, even for locals. Subcontractors were crucial cogs in the industrial employment machine; they often introduced relatives, friends, and neighbors to Linchang, who in turn recommended their own friends and relatives, resulting in a well-known pattern of chain migration. Most subcontractors and migrants we encountered were from Huize and Zhaotong counties, poorer and more rural parts of Yunnan. One middle-aged man from Huize, for instance, told us he first heard of Baocun from a relative who had married into the village. He introduced a young couple to his subcontractor, and in turn they introduced their young niece. One evening, the young couple invited us to their son's birthday party (see figure 3.4). We crammed into a small room with over a dozen other guests gathered around a low wooden table whose centerpieces were two identical white, frosted birthday cakes (one bought by us and one by the hosts). As we talked to some of the other guests, we realized most of them were relatives or neighbors in their natal village. The party in fact served as a useful introduction to several other migrants in the vicinity, who, given their long working hours, would have otherwise been difficult to meet.

Migrants' narratives of why they came to work in Baocun suggest the main drivers were push factors such as the poverty and remoteness of their hometowns and pull factors such as the opportunity to earn money to cover the costs of building a family home, children's education, and healthcare for the elderly. The tightly knit networks of support migrants built among them were not only useful for securing jobs, but also to fend off isolation created by the frequent discrimination they encountered. While their relationships with locals were not always inimical, stigmatization was common. Our host's mother told us not to walk along the railway line alone because "there are outsiders." One of the local doctors described the migrants as uneducated and dirty, and she blamed them for the high levels

**Figure 3.4**
Birthday party at a migrant's home, 2009.

of waste. Village cadres for their part despised some migrants for taking the opportunity to disobey the one-child policy and have two or three children.

Beyond such stigmas, migrants were structurally disadvantaged as compared to locally registered residents on several fronts. First, while some locals earned a considerable income from renting out as many as fourteen rooms to migrants, migrants had to cover the cost of renting accommodation. In 2009, the cost of renting a room ranged between sixty and one hundred yuan per month, for very basic accommodations. Rooms in the shantytown cost eighty yuan per month. Second, while locals were entitled to farmland contracted per capita (some of which they have used to build additional housing for migrants), migrants needed to rent land if they wished to engage in agriculture. Several migrants did so, in order to decrease their food costs. When migrants had a good relationship with their landlords, they might be allowed to farm their land for free. Uncle Chen, for instance, had come to Tacun from Huize County seven years previously and made thirty yuan working half-days (four hours) for

Linchang. He and his family rented a room from our host's sister for one hundred yuan a month and farmed two *mu* (1,333 square meters) of her land to feed his family. (These arrangements benefitted locals too, since they could only gain compensation for polluted farmland if it was not fallow.) Third, unlike locals, migrants were not entitled to receive either compensation for pollution or land rental fees, which in Tacun in 2008 were as high as 15,000 yuan per person. Fourth, their living costs were higher. In addition to rent, migrants were charged twenty yuan a month for "sanitation costs" (*weisheng fei*), and they paid higher rates than locals for water and electricity. Until 2007, the local school cost a steep 1,000 yuan per month, but since then fees have been reduced to 150 yuan per month. Fifth, given that migrants faced higher costs than locals and that the purpose of being in Baocun was to work, they often took on the most dangerous and strenuous jobs, and worked long shifts in order to earn higher wages. Perhaps most tellingly, only migrants died in a major accident in 2008, when a sulphuric acid depot exploded killing seven and injuring thirty-two migrant workers.

The desire to earn as much as possible and spend as little as possible caught migrants in a double pincer whereby they lived in poor sanitary conditions and were exposed to occupational health risks. Both of these conditions posed significant threats to their wellbeing in the short and in the long term, especially since many migrants stayed for several years, although such effects would be extremely hard to measure (Holdaway 2014). The worst accommodations were probably in the shantytown. It was very crowded, poorly built, and adjacent to the factory, and therefore most exposed to fumes and most at risk should an accident take place on the factory's premises. Rooms were small, kitchens were shared between several families, and public toilets were rarely cleaned (see figures 3.5 and 3.6). The following section discusses how those living in such conditions made sense of their lives and valued their hardship.

## Pollution and the Value of Migrants' Lives

In May 2009, we visited a family who had lived in the shantytown for seven years. The couple and their two children rented two rooms, of roughly nine square meters each. The main room was very clean and neat, featuring a double bed, a square table, a low table, a sofa, and a small cabinet with a

**Figure 3.5**
Migrant housing in the shantytown, 2009.

**Figure 3.6**
Interior of a migrant family's home, 2009.

television. The couple were from the same township as the labor subcontractor, located in Zhaotong County. The husband, Mr. Chen, served as a cadre in his natal village, but this only earned him one hundred yuan per month. Zhaotong is a poor county, and farming was not sufficient to cover school costs for their two children. In order to support his family, Mr. Chen took on not one but two jobs in Linchang. He earned 800 yuan per month for stoking the fire and eighty yuan per day for packing chemical products. When the factory operated well, he could earn more than 3,000 yuan per month. However, since the financial crisis, production had declined, and he only earned little more than 2,000 yuan a month. This barely covered their costs. Rent was 160 yuan (eighty yuan per room), in addition to water and gas fees. The main costs however were food: several migrants stated they spent 800 yuan per month on food for their families. Mr. Chen estimated that a couple without children needed 1,000 yuan per month, and those who brought children, as he did, needed an additional 500 yuan per child to cover food costs and schooling. He reflected that: "the salary is too low now, and the air pollution is serious. If nothing changes, I plan to go back to my hometown in a year or two."

Mr. Chen's case illustrates both the poor living conditions for migrant workers and the long days they are prepared to work to earn higher wages. Weiguo further elucidates this predicament. Weiguo rented a private room with his wife and two daughters, and he worked in Linchang. Like Mr. Chen, he was introduced to Linchang by a subcontractor from his natal county. He lifted and moved fifty-kilogram bags of fertilizer from a fork lift onto a conveyor belt, eight hours a day, twenty-three days a month. He estimated that he moved six to seven hundred metric tons of fertilizer per month, averaging at seventy bags per hour. This earned him three to four thousand yuan per month. Since he was paid by weight, he preferred not to wear gloves because they slowed down his work. But he argued that when bags were well packed, there was hardly any smell and that his work had little effect on his health. Weiguo had also done some lighter work, such as operating machinery alongside locals. But this would only earn him 600 to 700 yuan per month, which was not enough to support his family; therefore, he opted instead for the current work. His wife did "light work" in the fluorosilicic acid depot, which required wearing gloves, a mask, and protective glasses to decrease exposure to acid. Despite these occupational health threats, her pay was only 600 to 700 yuan per month.

Another example may illuminate the hazardous working conditions faced by migrants. A couple in their forties from Huize County rented a dusty and noisy room by the main road and farmed some of their host's land to grow some vegetables for their own consumption. They had seen footage of me helping an elderly lady lift some bags full of grain aired a few days previously on the local television channel. They were also related to a family I visited the previous night to celebrate their son's first birthday. They were therefore very welcoming to me and forthcoming in discussing their experiences in Baocun. They worked in Linchang loading raw material into processing equipment, making fifty yuan each for an eight-hour day. They were introduced to Linchang three years previously by a subcontractor who is originally from their village. Many of their relatives and neighbors were also in Baocun. Their daughter was married (and therefore no longer required their support), but they needed to raise funds for their son who was finishing high school and hoping to go to university.

When asked about their work, the husband replied bluntly: "The work we do is dangerous." He described the air in the workplace: "the smell irritates your nose, especially the new, fresh [raw] material." He was quite aware that this could affect workers' health. "One of my coworkers' teeth fell out without any pain. Maybe it is related to pollution at work. The air is so acid." However, when it came to pollution's effects on themselves, Mr. Zhang and his wife approached it with a mixture of self-assurance and daring wit. He proudly stated: "My health is good. I only ever had two injections in my life." He argued his own health was the reason he has not fallen sick like some of his colleagues. He also explained that he washed carefully every night, "otherwise my eyes will be red the following day." He conceded that this work was so harmful it may cause illness and death. He stated that people under forty were unwilling to take such work because "they are afraid of getting some serious and unknown illness." Indeed two young villagers who previously worked with raw materials told us they had quit this work because of its effects on health. But in the face of this possibility, Mr. Zhang quipped: "We're old enough. So we are not afraid of illness. The worst that can happen is death, so what?" His attitude encapsulates pride in his own good health, confidence that he was taking precautions, and despondence mixed with boldness in the face of death.

Phil Brown and Edwin Mikklesen (1997) found in their research on the infamous leukemia cluster in Woburn (Massachusetts, USA) that many locals preferred to pretend the problem did not exist, because it was too painful to contemplate that they were faced with a threat they were scarcely equipped to avoid. Mr. Zhang's attitude resonates with this description to the extent that he stressed that he was healthy enough and careful enough to overcome the threat of pollution. This may be his way of reassuring himself that he could at least minimize the risks. Thinking constantly about the dangers posed by working and living in a severely polluted environment would be emotionally exhausting. It would add to the physical burden of pollution the psychological burden of fear, dread, and hopelessness. It is much easier to face these circumstances if one tells himself or herself that the circumstances are not so bad after all. But the attitude shared by Mr. Zhang and his wife goes a step further in acknowledging and accepting the possibility of death with seemingly relaxed resignation.

Can Mr. Zhang and his wife really place so little value on their lives? This question goes to the core of the experience of living and working in Baocun, particularly for migrants, but also for other low paid and relatively poor locals. The answer depends on some deeper, more analytical, and more pertinent questions: *How* do they value life? What do they think their lives (and deaths) are worth? Italian philosopher Giorgio Agamben (1998, 2005) has argued that bare life (*zoe*) characterizes those—such as illegal migrants, asylum seekers, and Guantanamo Bay prisoners—who have been stripped of their legal rights and their citizenship. As a consequence, they only exist as biological life; their existence is politically insignificant. Migrants to Baocun are also constituted as bare life to the extent that they have limited rights and entitlements, given that their formal residence is elsewhere. In their case, the denial of rights is not only exercised by the state, but rather by a political-economic nexus of local government bent on raising its own revenue, exploitative firms that prioritize profit at the expense of workers' welfare and environmental protection, and, last but not least, workers who have internalized and by and large accepted their powerlessness. Under such circumstances, not only are they relegated to the status of bare life, but their bare life itself is under attack as they engage in harmful, polluting work and live in insalubrious conditions.

Baocun migrants' experiences, however, also differ from Agamben's understanding of bare life in one important respect. Where he emphasizes that the essence of modern sovereignty is to have power to strip lives of social and political value, migrants still regard their lives as valuable, despite their political and economic marginality. They may be excluded from legal rights, have limited welfare, and work and live in appalling conditions; but their willingness to face such hardships and such attacks on both their political and their bare life paradoxically endows their lives with meaning and value. In the face of the little consideration for their lives by the factory and the (local) government, migrants reclaim a moral value for their work: the value of supporting their families (see also Lora-Wainwright 2009, 2013a). Indeed, most often, migrants such as Mr. Chen and Mr. Zhang explained that the main purpose of seeking waged work away from their home villages is to provide for their family. A frequent refrain among migrants was "we have a heavy burden, we have young [children] and old [parents] to care for." In this sense, they valued their own lives, and particularly their labor power, for their potential to support their family.

The physical cost of engaging in such work was understood to be so severe that it was deemed inappropriate for those who had not yet laid the most important foundation of family life: marriage and childbearing. Chunyan, a newly married woman in her early twenties, recalled that before her marriage she worked alongside some of her friends packing and loading chemicals in Linchang and had frequent nosebleeds as a consequence. She reasoned that she stopped because she was still single, but that it did not matter to her friends, who had already married. She explained: "If you have not married and had a child you should stay away from poison" (May 20, 2009). This betrays a conception of women's healthy bodies as a form of capital needed for finding a husband and bearing children, and which can subsequently be sacrificed for their benefit. In turn, it implies that one's own life (particularly a woman's life) is valuable to the extent that it serves to create, and later support, a family. Men's sexual and reproductive capacities were also threatened by such hard labor. Rumor had it for instance that a man who worked in Linchang had become impotent and had bought his wife a dildo. Couples then may be rendered incapable of enjoying one of the pleasures of married life by the necessity to provide for their family through work. Factory work stands in a paradoxical relationship to procreation: it threatens the capacity to have children, but it is required to sustain the family once children have been born. Such limited value placed on one's health per se, seeing it instead at the service of one's family, could be taken as an example of the much debated centrality of the extended family in Chinese culture and the lack of focus on individuals, though recent research suggests this is not (or at least no longer) the case (Yan 2003, 2009). While the value attributed to the family in Chinese culture may indeed be a factor, it is still necessary to ask why it should manifest itself in such extreme ways, and what is specific about this context. Such erasure of the value of individuals' health for its own sake is a telling sign of the difficult choices migrants and poor local residents face and of how powerless they feel.

Against such a grim scenario, it is important to note that not all are ready to sacrifice their wellbeing or to let close relatives do so. Mr. Chen (cited above), who lived in the shantytown, said he would return home unless his income rose because pollution was too severe. Meiting, a woman in her late thirties from Huize County, had already lived in Tacun for six years with her husband and one of their two children. She was employed by Linchang doing light work and her husband was a construction worker. She reflected that laborers might earn up to 5,000 yuan working in Linchang, but that the work was tiring and straining. Referring to the case of a man who had become impotent, she lamented: "You cannot trade life for money." Some migrants also presented alternative narratives to make sense of their choice to engage in heavy labor. Meiting explained that she and her husband had left their hometown because: "The village only has a dirt road. We can't even sell vegetables. We feed what we can't finish to the pigs, and we can't even sell those. We just slaughter them and preserve the meat for the family to eat." Baocun by contrast offered opportunities to earn a living and support children's education. Similarly, Weiguo described his life and work in Baocun as a choice he made after ruling out several other alternatives. He had previously worked in a shoe factory in Zhejiang province, but he quit because he could not bear the heat. He quit construction work for the same reason. Work in Linchang was strenuous, but he preferred it to his previous occupations because it was relatively well paid. He also preferred to be in Baocun than to be at home: "Life in my hometown is boring. I like to come out for work. I go to work every day and have a salary every month. When I'm free, I can watch TV, go out and have fun, smoke a cigarette. I prefer life here. I don't even want to go back." Many of the young migrants I came to know during my previous fieldwork in Sichuan made very similar statements upon returning home for Chinese New Year: migrating for work is tiring, but being in the village is boring and makes it hard to earn money.

In light of these seeming glimmers of hope and agency, it is worth returning to the question of how migrants value their lives. Indeed, the fact that not all are prepared to sacrifice their health highlights even more painfully the predicament of those who do. That some, like Weiguo, would present their work in Linchang as a choice is itself a symptom of their powerlessness and abjection. And the fact that migrants like Mr. Zhang and his wife seem to accept the likelihood of an untimely death is an even deeper injustice than the mere fact that they engage in such harmful work. Their resignation that they cannot aspire to better, healthier work bespeaks their vulnerable position within an industrial system that thrives by exploiting and harming its labor force while wreaking havoc on the environment. The low value they place on their own health is derived at least in part from the little regard their employers and local bureaucrats and regulators have for them.

This account illustrates some of the less visible and embodied ways in which toxic natures become part of everyday life, how this affects attitudes toward pollution, and the life choices ambivalently embraced while accommodating pollution. These processes powerfully mold the subjectivities of those at the bottom of the ladder and the forms that resigned activism may take. In the case of these most disadvantaged socioeconomic groups, it largely amounts to a deep-seated sense that their health is subordinate to the welfare of one's family, interspersed with moments when the most harmful tasks are refused in view of seeking a healthier life. Next, I turn to how these perspectives shape the forms of activism present among this socioeconomic group.

### Precarity, "Disaffective Labor," and Slow Violence

The term "precariat" (a fusion of "precarious" and "proletariat") has risen to prominence recently to describe "an emerging class characterized by chronic insecurity, detached from old norms of labour and the working class" (Standing 2014, 1). For Guy Standing (2014), the precariat is unlike the proletariat in so far as it encompasses middle-class white-collar workers and others who are not, strictly speaking, poor, but are part of a flexible and disposable temporary workforce. To the extent that Baocun migrants have precarious employment and rights, the term indeed applies to them. Precarity in their case also extends to their health: their lives themselves are precarious, threatened as they are by both accidents and by long-term exposure to toxic substances. The term precarity is not new, but rather it is situated within a longer genealogy of debates about marginality, informality, and social exclusion (Munck 2013). What sets the precariat apart in Standing's writing is its potential to become a "dangerous class" that may come to threaten the ruling classes. On this front, Baocun migrants and the precariat part ways. Indeed, the precarious nature of migrants' position in the workplace and in the local community denies them any sense of entitlement, even to demand better working and living conditions; they have by and large internalized their sense of powerlessness. After all, as many of them put it, it was their choice to come here; if they make too many demands they may just be laid off, and someone else will be readily found to replace them. Baocun migrants are what Kevin Bales (2012), in the context of diverse forms of contemporary global slavery, calls "disposable people." They are disposable to the extent that they lack job security, but also in the sense that their health comes to be a disposable good in the service of capital accumulation.

These conclusions about migrants' powerlessness and resignation may seem in stark contrast to recent scholarship that documented growing numbers of labor protests in the past decade (see for instance C. Chan 2010; C. Chan and Pun 2009; J. Chan and Selden 2014; J. Chan, Pun, and Selden 2013; Lee 2007; Pun and Lu 2010). Several studies suggest that the profile of migrant workers has changed from a first generation who joined the labor market in the 1990s and returned to their villages to marry, settle, and raise their children (see R. Murphy 2002, 2009; Pun 2005; Rofel 1999) and a second generation of younger workers (born in the late 70s and 80s), with higher skills, education, and aspirations to settle in urban environments and enjoy more comprehensive welfare (Chan and Selden 2014). Some have argued that members of the second generation are undergoing a process of proletarianization, which involves shared grievances and solidarity fostered by working and living together in factory dormitories (C. Chan and Pun 2009). They have become a "semi-proletariat" who demand pay raises, better conditions, and welfare benefits (J. Chan and Selden 2014; Chan, Pun, and Selden 2013). They are increasingly aware of their class position, and their capacity for collective action is improving (Pun and Lu 2010). Yet even these scholars concede that protests have led to little or no change on the part of industry (Chan and Selden 2014). Even less optimistically, A. Chan and K. Siu (2012) argue that there is so far no evidence of large-scale, coordinated protests between workplaces that would demonstrate the presence of a shared class consciousness among migrants.

Baocun's shantytown, which accommodates migrant workers, comes as a close, horizontal approximation of the factory dormitory blocks that C. Chan and Pun (2009) see as the backbone of proletarianization. However, there was little or no evidence of solidarity between workers coalescing into protests to demand better working conditions or better wages. Even when they faced extreme events, such as the 2008 explosion that left seven dead and thirty-two injured, migrants seemed to react on a family-by-family basis, rather than by mounting protests. The lack of solidarity and organized action among Baocun migrants may be due to the different historical and political economic profile of localities in inland China like Baocun as compared to those of the Pearl River Delta's special economic zone observed by J. Chan, N. Pun, and M. Selden (2013). At the most basic level, in Baocun, wages are considerably lower and the supply of older, less demanding laborers relatively plentiful (cf. Davis 2014). The demographic profile of migrants is also different from those belonging to the second generation described above: younger, more skilled, and aspiring to settle in the city. Studies have suggested that, as of 2013, 46.6 percent of China's 269 million internal migrant workers were born after 1980 (Chan and Selden 2014, 600) and most were unmarried (Chan and Pun 2010). Migrants to Baocun by contrast were rarely younger than thirty and almost invariably married. Perhaps for this reason, their expectations and aspirations were significantly lower than those of the young migrants described by Chan, Pun, and Selden. Indeed, if Mr. Zhang's and Chunyan's statements are to be taken at face value, only those who are older and already have children are likely to come to such a "dangerous," "harmful" place.

Disenfranchised as they are, migrants to Baocun are also disillusioned and disaffected: they lack faith in any potential to better their own condition by uniting and making demands of their employers. Whereas for Marx estranged labor and alienation were at the root of rising class consciousness and solidarity among workers, in Baocun they seem to have had the opposite effect: not the empowerment of workers to demand better rights but rather a resignation to the status quo. Indeed, the migrant population of Baocun may be a class-in-itself (they have a shared relationship to the means of production), but it is not a Marxian class-for-itself, to the extent that the workers do not organize to pursue common interests. Instead, migrants try to ignore or minimize threats to health discursively (reflected in common statements such as "it's best not to think [about the dangers]"), and physically, by closing windows at night, wearing masks, and washing carefully after work.

Writing on affective labor as an aspect of immaterial labor, M. Hardt (1999, 89) argued: "laboring practices produce collective subjectivities, produce sociality, and ultimately produce society itself" (see also Lazzarato 1996). Affective labor is "human contact and interaction," "a binding element" (Hardt 1999, 95) that produces social networks, forms of community, biopower; "its products are intangible: a feeling of ease, wellbeing, satisfaction, excitement, passion—even a sense of connectedness or community" (96). He argues such labor is not new (feminist analyses for instance noted its importance in kin work and maternal activities), but what is new is its increasing value in capitalist accumulation, and conversely its potential for liberation (100). Work in Baocun's industries by contrast amounts to a form of what I would call dis-affective labor: rather than creating connections and community, it undermines them. By virtue of their marginal and vulnerable structural position, work in Baocun isolates and exploits migrants. Although they work alongside locals, they do not join forces to demand better working conditions. Locals' demands to the local firms focus on pollution compensation, demands that migrants who are not formal residents of Baocun have no right to join. As I explained above, locals receive land rental fees from mines, regular and ad hoc compensation for pollution; they may rent out their land and accommodations, and they are typically better positioned to establish private businesses. The different entitlements and opportunities open to villagers versus migrants undermines the formation of a shared class consciousness. More broadly, the hugely uneven structural positions of formal and informal employees leaves little room for solidarity to develop across the divide.

In 2010, eighteen young employees of Foxconn, the world's largest contract electronics manufacturer (which provides products for Apple, among others), attempted suicide by jumping from high buildings on the factory premises in Shenzhen, resulting in fourteen deaths (Chan and Pun 2010; Chan 2013). For Chan and Pun (2010), Foxconn suicides may be regarded as a form of protest against an exploitative, inhuman labor regime, which sacrifices human dignity for corporate profit. These young workers "reject the regimented hardships their predecessors endured as cheap labor and second-class citizens. They rebel against their marginalized status and meaningless life" (29). Might the much slower, protracted, and less visible suffering of Baocun migrants also be seen as a sacrifice?

Baocun migrants are demographically and politically unlike Foxconn's workers: they are older, married, and largely resigned to their plight. Their suffering is not as visible as that of Foxconn workers, nor is it manifested in such extreme measures as suicide, but it is nevertheless a form of sacrifice enacted by an exploitative labor regime. The key differences lie in the diverse temporality and visibility of these two forms of suffering. Foxconn's "suicide express," as it was termed, was an acute response to unbearable working and living conditions that attracted immediate attention and global media response (if less than satisfactory corporate redress). Baocun migrants' slow, day-in, day-out exposure to pollution may also result in untimely deaths, and migrants are increasingly aware of this (as Mr. Zhang testifies). But such deaths are doomed to remain invisible for several reasons. Long latency periods and complex causality for many illnesses make it difficult, if not impossible, to establish the extent to which pollution precipitated illness. The fact that migrants often move between occupations, and ultimately may even move back to their hometowns, also complicates the task of linking exposure to pollution with illness (see Holdaway 2014).

The Foxconn suicides befit traditional definitions of violence as "an event or action that is immediate in time, explosive, and spectacular in space, and therefore eminently visible" (Nixon 2011, 2). By contrast, Baocun migrants are victims of "slow violence": "a violence that occurs gradually and out of sight, a violence of delayed destruction that is dispersed across time and space, an attritional violence that is typically not viewed as violence at all" (2). Slow violence is "incremental and accretive, its calamitous repercussions playing out across a range of temporal scales," and therefore relatively invisible (2). Such invisibility adds to the suffering of its victims. Their invisibility is—as much for Baocun migrants as for Foxconn workers—an effect of "structural violence." Nixon draws on this concept, originally elaborated by Galtung (1969) to describe "structures which can give rise to acts of personal violence and constitute forms of violence in and of themselves" (Nixon 2011, 10). The term has also been adopted in critical medical anthropology to refer to the uneven burden of disease for those at the bottom of the social ladder, and to the ways in which poverty, marginality, and illness are spatially mapped onto each other (see Farmer 2003). For Baocun migrants, the structural violence of their marginal and vulnerable position is compounded by the slow violence of their exposure to pollution, likely to cause long-term health effects and yet elusive to demonstrate. These forms of violence create a deep-seated sense of resignation to pollution among migrants. This, in turn, results in the absence of collective action among migrants, despite their clear awareness of the harmfulness of pollution.

## The Origins and Effects of Resigned Activism

Like migrants, local residents were also acutely aware of the potential harm of pollution. However, as I highlighted earlier in the chapter, locally registered residents were a very diverse group, with uneven opportunities to avoid pollution. Indeed, those who were able to move away from Baocun, such as local elites, had already done so. Many of those left behind voiced frustration at their limited access to financial resources required to buy a home elsewhere. Slow and structural violence affected those at the bottom of the socioeconomic ladder and fostered feelings of resignation. A middle-aged woman quoted in the opening passage of chapter 1 phrased such resignation most succinctly: "I live as long as fate allows" (*huo jitian suan jitian*).

While this statement resembles closely the sense of resignation prevalent among migrants, locally registered residents—even those at the bottom of the socioeconomic ladder—were structurally advantaged and could resort to a greater range of alternative livelihood options. As a consequence, resignation among poor locals had different origins and took on different connotations than it did among migrants. Above all, it intersected more clearly with resilient, adapted forms of activism, which lay largely beyond the reach and (self-)entitlements of migrants. In part, resigned activism among poor locals was shaped by a sense that they, too, benefitted from industrialization through work opportunities and compensation packages, and that pollution was an inevitable side-effect of development. A nineteen-year-old local factory worker, for instance, explained that factories cause harm, but this is the price to pay for work. A woman in her twenties similarly stated: "It's normal to have pollution in a village with mines and factories. ... People received rent from selling their land, so they should bear pollution." Just as for migrants, working in polluting firms acted as a form of disaffective labor, disempowering villagers from demanding less pollution.

Poor locals' attitudes to pollution and their responses were slowly molded through decades of experience interacting with polluting firms and with different generations of local officials. This evolution can be mapped across three phases of growing symbiosis between pollution and the locality.[5] Pollution first became inescapably apparent to locals in the early 1980s: cows' legs swelled and they perished soon after; and industrial wastewater dumped in the local irrigation killed crops and caused many to develop "strange illnesses" (illnesses they had rarely heard of before), such as gall stones, and pain and swelling in their hands and feet (possibly symptoms of fluorosis). Villagers sought the support of higher levels of government, and they compiled a petition to the township, complaining that Linchang (a state-owned and state-run firm at that time) did not care about the local community and the environment. The petition was rejected, however. Village officials were allegedly bribed by the industry and locals were pressured to abandon their demands.

This experience taught locals that their activist strategies were ineffective and their expectations were unrealistic. It ushered in a second phase (spanning the 1990s and early 2000s) in which the language of caring for the community was largely abandoned and the value of the environment was reframed in more tangible, materialistic terms: demands for compensation for damages incurred. Locals also learned to keep their complaints low-key and localized, mostly designed to raise the alarm among local firms and mobilize village officials to negotiate on their behalf. They organized blockades on the occasion of acute instances of pollution, such as acid leaks, explosions, or particularly pungent air pollution. However, village officials at this stage were mostly unsupportive. Things changed significantly when a new generation of village leaders was elected in 2004. These younger cadres were experienced with industry and positioned themselves as mediators between the industries and villagers, securing much more significant compensation packages and land rental fees than had been available in the past. Growing financial benefits for locals during this third phase further entrenched economistic ways of valuing the environment and rendered pollution seemingly inevitable. It caught villagers in a "compensation trap" (Van Rooij, Lora-Wainwright, Wu, and Zhang 2012): compensation deals determined how villagers' related to the local environment and to their lives more widely, rendering other demands (for a healthier environment) infeasible.

Local officials, in collusion with polluting firms, further nourished disaffection by offering targeted compensation packages that in turn silenced any potential complaints. Over the course of three decades of liaising with industries and local officials, villagers learned to accept pollution as a fact of life. For the local government and industry bosses alike, encouraging resignation was a way of condoning pollution and affirming it as part of Baocun's status quo. Local officials acted in ways similar to what Mary Gallagher (2014), in the context of labor disputes, described as "the activist state": they made use of their power and connections to act as intermediaries between villagers and local firms and to mitigate discontent through compensation deals, thereby avoiding complaints to higher levels of government. For their part, some industries adjusted their operations to fit with local needs. One small firm, for instance, paused production during the first and most productive farming season (during which produce also sells for a higher cost) to avoid affecting the crops. This of course also meant that they could not be held liable to offer compensation should the produce be substandard. It is a powerful example of how local firms—particularly smaller ones—responded to villagers' demands, while at the same time influencing villagers' sense of what demands might be acceptable and productive.

Over time, locals' experiences with various ways of engaging with industry shaped the ways in which they regarded pollution, and their expectations not only about the environment, but also about their health. Their activism took on a much more economistic and individualistic tone. This was in line with their redefinition of the environment as a good that can be compensated for, rooted in social divisions and stratification whereby locals focus on obtaining compensation for damages and on securing welfare on an individual and family basis.

Through these various negotiations and adaptation processes, pollution has become part of Baocun's natural environment; nature is toxic and toxicity is natural. In her study of NASA's astronauts and the USA's space biomedicine program, Valerie Olson (2010) argued that in "extreme environments" such as outer space, doctors apply the concept of "space normal" to refer to bodily and environmental conditions that may be pathological on earth, but are regarded as normal in space (see also Saxton 2014). Similarly, in Baocun's extreme environment, villagers came to accept that different parameters applied there than elsewhere. Accepting and normalizing common health effects of pollution were part and parcel of this process of resignation. Villagers adjusted their standards of health to accommodate conditions such as nose and throat infections, hand and feet swelling, and joint pain (the swelling and pain, again, possibly the early stages of fluorosis) that are endemic and closely correlated with types of pollution prevalent in Baocun. Illnesses that used to be regarded as "strange" have now become familiar. A young man who had suffered with recurrent nose infections for the past six years explained that his doctor told him that this was "normal" in a polluted place like Baocun. He quipped: "These are small problems; once you're used to them you're fine." This normalization ran so deep that, when asked about their health, villagers almost always neglected to mention common (pollution-induced) endemic problems such as nose infections. Indeed these conditions, like pollution more broadly, had already come to be regarded as part of life.[6] Most poignantly perhaps, some locals argued that bodies themselves can become accustomed to pollution and therefore be less affected by it. Perhaps they hoped that if they psychologically and emotionally adapted to pollution, bodily adaptation would follow too. Whether they genuinely believed this to be the case or whether this was yet another coping strategy to deny the severity of pollution's effects is hard to determine. What is certain is that it articulates a clear sense of interdependence between the new natural/normal and locals' bodies. It goes hand in hand with a revised and resigned mode of activism designed to adapt to pollution's presence rather than to question it or to demand better living and working conditions.

Guo Lin's case clearly illustrates the attitudes and responses of poor locals toward industry and pollution. Guo Lin was a thirty-one-year-old man, born and bred in Tacun. Yunmei (one of the fieldworkers) and I had the good fortune to meet him on May 12, 2009 as he walked home, and he invited us to join him. His house was modestly sized. The building was relatively old, but the interior was redecorated with bright white floor tiles.

The central room featured a colorful poster picture of twin babies surrounded by abundant fresh fruit, a sofa made of polished wood, a low wooden table, and some small plastic stools. Yunmei and I headed for the plastic stools—a way of displaying modesty and showing humble gratitude for the hospitality, something I had learned to do when I lived in rural Sichuan—but Guo Lin insisted we should sit on the sofa and offered us some hot water for our portable cups. He told us his family had two *mu* (1,333 square meters) of land, but planting crops was no use: once the thick dust floating in the air settled on them, the plants no longer grew. As a consequence, his family (he, his wife, and their young son) relied almost exclusively on his 1,000 yuan monthly income as an informal worker at Linchang. As a crane driver, he worked for twelve hours a day, in cycles of twelve days on and five days off. This earned him just enough to cover his family's expenses. Some years ago, he used to do what he referred to as "poisonous, harmful work" (*youdu youhaide gongzuo*), which involved close contact with raw materials. "I didn't want to do that job, but I had just got married, and had a lot of financial pressure," he explained. This work was much better paid than his current job, but he believed it caused him a severe lung infection and chronic cough. He eventually quit the job, but was apprehensive about finding another one. Even for local villagers, finding work at Linchang that is not strenuous, poisonous, or harmful requires good "connections" (*guanxi*).

Guo Lin was acutely aware of the harm of pollution on crops, livestock, and locals' health. He had taught himself some basic principles of Chinese medicine and was one of only a few local men who did not smoke. He also used to enjoy watching investigative journalism programs on TV. His vivid description of local pollution drew on several examples: many of the local trees had died because of air pollution; crops fail routinely, and recurrent acid spills kill anything that manages to grow; the water is polluted beyond remediation (villagers paid for independent tests, which found it to contain fluorine and mercury); cancers of the respiratory and digestive tract have increased (he estimated more than thirty cases in Tacun, the subvillage closest to Linchang); many locals have crooked teeth (a symptom of fluorosis); the incidence of kidney and gall stones is as high as 40 percent; lifespan has decreased; and children are less healthy than those from elsewhere. He believed his son, for instance, was shorter than average, as if he was one or two years younger (see Lora-Wainwright 2013d).

The culprit for such extensive damage was not just Linchang, but several other local fertilizer plants, mines, and related businesses. Small protests and blockades, he explained, were a routine response to frequent accidents. One of the most severe was an acid leak in 2005 caused by one of the local firms. When the firm refused to engage in negotiations with villagers, they blockaded the firm's gates. Most of those attending were old ladies, specified Guo Lin, because the police were less likely to arrest them and because their jobs were not on the line (although of course their relatives' jobs might be). Eventually they received some compensation for the damages. But Guo Lin was skeptical that these smaller firms, or even bigger plants like Linchang, really cared about the environment. "Linchang are better at environmental protection now, but they pay attention only when they are under pressure. If they cared, they would have installed proper equipment in the 80s and 90s." Reflecting on television reports of other polluted villages, he commented: "Here it's not like those cases on television, where people complain and problems are solved." He believed this was due to the size and economic importance of local industries, particularly Linchang. "We can do nothing about it, the factory is a big tax earner for government. It's the same all over this township. We sacrifice the 'small self' for 'the big self' [the nation]." In view of this, despite his clear awareness of the dangers of pollution he claimed he preferred not to think about these problems: "If the eyes don't see, the heart doesn't worry" (*yan bujian, xin bufan*); it is better "not to think about things one cannot change" (*bu neng gaibian jiu bu qu xiang*). Indeed, Guo Lin believed it would be impossible for the village to clean up: "To change you need enough money to move, change the air, water, and soil," but his family, like most others, did not have the resources to do this. All he could do was avoid the most harmful jobs.

Guo Lin's account articulates an activist resignation to the fact that industry inevitably entails pollution. His choice "not to think about things you cannot change" resembles migrant Mr. Zhang's refusal to ponder the full extent of the potential effects of pollution on his health. There are some important differences however. First, Guo Lin also voiced deep resentment toward local industries that ruined their surroundings, the local officials who failed to protect them, and the political economic calculus whereby the welfare of their locality was deemed disposable for the benefit of national development. This betrays a resilient sense that a better environment *should* be possible and that the local government *should* be doing more to protect it. Such regret among locals about the current state of the village may also be rooted in their memories before full-blown industrialization. They painted vivid pictures of forests populated with wild animals, fertile land, and crystal-clear and plentiful streams where they swam and washed. Second, for migrants resignation meant keeping one's head down and accepting the risks without any prospects of demanding improved conditions. For locals, by contrast, it still left open the option of requesting compensation. Although this form of activism had been disciplined and molded to fit with new parameters of what was regarded as possible in Baocun, it still offered an avenue for demanding redress. Third, and perhaps most important, unlike most migrants, Guo Lin did quit dangerous work in favor of a lower paid, but less harmful, job once he had the opportunity to do so. Like migrants, however, Guo Lin and other locals remained keenly aware that the odds were stacked against them, and the stakes for anyone wishing to stand up to polluting firms were extremely high. Put most poignantly, in the words of a local woman in her thirties who frequently played a leading role in local blockades: "The wealthy fear those who do not fear death." By implication, we might conclude that those daring to oppose wealthy firms risk their lives, and those who are not prepared to put their lives on the line should resign themselves to the presence of pollution and target their activist efforts and their expectations accordingly.

## Conclusion: Resigned Activism and the Value of a "Good Life"

In the wake of the 2010 Foxconn suicides, dozens of demonstrators protested outside the company's headquarters in Taipei, holding banners that read: "What is the price of a human body?" (*xue rou he jia*) (Chan and Pun 2010, 23). How do poor Baocun villagers and migrants value their own lives in the face of persistent structural and slow violence? An anthropological approach demands that we pay attention to how parameters for what constitutes a good life, or even an adequate life, are shaped, not only by the structural conditions dictated by political economy, limited labor rights, and environmental protection, but also by the subjects themselves. Locals (and even migrants) regarded themselves to be in part beneficiaries of polluting industries, to the extent that the industries offered them waged work. In view of this, they were prepared to compromise significantly over their health and the state of the environment. Resignation to pollution and even suggestions that one's life itself was a disposable good, however, do not preclude alternative ways of valuing life, endowing difficult choices with moral connotations.

June Nash (1979) described the codependency of mines and miners in Bolivia with a memorable quote: "We eat the mines, the mines eat us." Similarly, in Baocun, the relationship between workers and local firms was one of mutual dependence. We must of course question the local, national, and global capitalist system that demands some workers should sacrifice their wellbeing and some localities sacrifice their environment. Much evidence about the dire working conditions and environmental health damage inflicted upon workers may encourage us to conclude that informal workers—migrants and locals—are overwhelmingly the victims of industrialization. However, if their voices are to be heard and their experiences to be understood in all their facets, we must also examine how they value their own lives, and consider that, from their point of view, enduring these hardships is the best they can do, for themselves and for their families. Putting such resignation under scrutiny, and acknowledging the limited, strategically framed activism among workers, are surely the first steps toward not only understanding their plight, but also demanding that it should stop.

# 4 "Undermining" Environmental Health: "Madness," Struggles for Clean Water, and the Challenges of Intervention

## Introduction

The following excerpt is taken from my fieldnotes of August 31, 2012:[1]

Today was a lesson in rejections. My assistant and I had talked briefly to Fengcun's subvillage head yesterday and he told us he would see us this morning. We got up early and walked over a mile from our host's home, reaching the small teahouse his wife runs at 8 a.m. He was nowhere to be seen. We politely inquired with his wife and she told us he had set off in the very early morning to go fishing. The reason for his refusal to engage with us remains in the realm of speculation. Maybe he really did go fishing. Maybe he had no intention of speaking to us but could not tell us directly, fearing a "loss of face" (something particularly undesirable in Chinese culture). Maybe he actually wanted to meet us but was told by his superiors not to do so. Reluctance to engage with researchers understandably seems a relatively common response when the topic at hand is sensitive. My identity as a foreigner no doubt considerably increases the likelihood of this sort of response, and it is a hurdle which never ceased to trouble my efforts to research rural pollution and health.

All in all, this fieldtrip has been quite taxing. Villagers who had already met me in 2010 during the first round of fieldwork were very welcoming, but my assistant and I have also faced a lot of rejections and frustration. I would never have imagined that when I greet villagers along the road they would ignore me. This never happened to me anywhere else in rural China. People are usually at least intrigued enough to grunt a bemused response to a foreigner who travels to what seems to them such an unlikely place. But here, not even a grunt. Some refused to speak to us because they said they had already interacted with enough researchers. Others complained that they've had tests on their blood, their hair, the local water and crops over the years [some of which were done by medical geographers who initiated work in this site, see below] but haven't seen any results. [This complaint is at least partly inaccurate. Indeed, the results of some tests were in fact conveyed to those who took part, and attempts to design public health education materials together with the leaders of the county Center for Disease Control (CDC) are ongoing, as I explain below.] Their frustrations are further fueled by the recurrent visits by environmental protection bureau staff and other county officials, without any visible outcomes. It particularly irritates them to see officials "coming here, not even talking to us, just taking some water, and leaving again," as many complained. Who can blame them for sometimes being despondent towards us?

I sympathize with their frustrations, but it also frustrates me to feel I failed to present myself as someone who actually wants to help. It hurts particularly badly, because I do want to help. Of course, I make no claim whatsoever to being in a position to help them effectively. My only hope to achieve anything of use to the local population relies on understanding their own position better and discussing options with other project team members and with those who are best positioned to act: local CDC staff. But the situation is so complex, the topic so sensitive and possibilities to carry out fieldwork so limited, especially for a foreigner like myself, that I feel severely curtailed in my ability to produce useful knowledge.

At root, the refusal by some to discuss their difficult positions and experiences is linked with a sense of helplessness. Those who spoke to me suggested that most often they failed to get the redress they sought, so they lost hope. But this helplessness is also a way to accept the status quo. Not only do they suffer the environmental health consequences of mining, but they also concluded they cannot live any other way. The only villager [Li Fang] who persistently tried for decades to gain redress was declared "mad." The extent to which he really was mentally ill also remains open to speculation. What is clear is that he refused to accept his powerless position. After only a few weeks in Qiancun [the administrative village of which Fengcun is part], I can't imagine how anyone could find the strength, determination, energy, and time to keep demanding something better. Those who do, have my utmost admiration. Those who don't, have my full sympathies.

This chapter looks at life in the shadow of lead mining in Qiancun village. It probes the development of resigned activism, feelings of helplessness, and relative ambivalence toward research activities. It examines locals' shifting relationship to mining, their frequent petitioning efforts, and demands for water tests and safe drinking water. Finally, it considers the challenges faced by the wider project of which my research was part.

### Qiancun's Lead and Zinc Mine: Socioeconomic Stratification, Environmental Health Impacts, and Villagers' Responses

Qiancun is a mountainous village located in Fenghuang, a county situated in western Hunan province (Xiangxi) in central China. Fenghuang was classified as a poor county (*pinkun xian*) in 1986, and in 2002 it became one of the National Poverty Alleviation and Development Key Counties (The

State Council Information Office of the People's Republic of China 2002). Qiancun in turn is identified as a poor village (*pinkun cun*) with relatively low agricultural potential. This makes lead mining and related processing an attractive livelihood option in comparison to agriculture or migration, with the potential to offer an income six times as high as agriculture (Ran 2012). Qiancun is an administrative village composed of four subvillages, located in a river valley surrounded by terracing. In 2010, there were 346 households and 1,560 registered residents, of whom about 560 were of working age (between sixteen and sixty-four). Almost all male residents in Qiancun share the same family name, Li. Fengcun is a subvillage with a local population of roughly 500. It is located downstream from mining and processing activities, and therefore the effects of local mining and processing of lead and zinc are particularly prominent. Most of my own research focused on Fengcun, while other members of the wider multidisciplinary project collected data across all four subvillages.

The Ministry of Land and Resources (2012) describes Hunan province as "a land flowing with non-ferrous metals," which accounts for one-fifth of total lead-zinc production in China. The Xiangxi region contributes most to this (Ministry of Land and Resources 2004). The development of mining in Qiancun village may be divided into five phases—elaborated in full in a previous article (Lu and Lora-Wainwright 2014)—which roughly correspond to the development of China's mining industry more widely, its policy context and political-economic shifts, though timelines and turning points may differ (see Wright 2011). In phase one, until the late 1950s, mining was limited to primitive techniques, mostly involving hammers and simple tools. In phase two (the late 1950s through the late 1970s) a state-planned economy was established and mining rose to prominence. In line with this, Fenghuang County authorities laid claim to mineral deposits on behalf of the nation, medium-scale mining operations were initiated in Qiancun by a state-owned mine, and a processing plant was opened in the 1970s with a capacity of thirty tons (see figures 4.1 and 4.2).

The institutional context of a planned economy and collective ownership of land meant that mining was largely monopolized by the state.[2] Therefore, no compensation was given to the village for loss of land at that time, nor was any provision made for the negative effects mining might have on rural livelihoods, or to ensure that local communities were able to enjoy the benefits of mining. Although more than two hundred workers

**Figure 4.1**
Part of the former state-owned mine in Qiancun, 2010.

**Figure 4.2**
A tailing pond resulting from decades of mineral processing, 2010. Fengcun village is to the right, below the tailing pond.

were recruited by the state-owned mine as full-time workers earning fixed wages, there were virtually no locals among them. Villagers resented the lack of employment opportunities and loss of their land, and demanded the right to mine. In 1972, an agreement was reached—still valid at the present time—between the state-owned mine and representatives from the village and the township government (respectively called the "brigade" and the "people's commune" at that time) to clearly mark the state-owned mine precinct with a "red line" and to allow villagers to mine outside it, in areas with lower quality deposits.

Phase three (late 1970s to early 1990s) involved continued mining activities by the state-owned firm, but, with the start of a period of economic liberalization in 1978, township and village enterprises (TVEs) were allowed to open (including mines) in order to absorb rural labor surplus and launch rural economic development (Gunson and Jian 2001; Kanbur and Zhang 2009; Tilt 2010). The encouragement to rely on local natural resources to fuel local development, national demand for lead, and aspirations to export underpinned the opening of a growing number of smaller and less regulated mines, some private and some collectively owned (at the village or township level). Indeed, several lead-zinc and mercury mines and smelting plants were established in Fenghuang at this time. As a consequence, during this phase villagers were increasingly in a position to benefit from mining, although benefits were largely limited to political and economic elites (i.e., village and subvillage leaders and some richer households), who were in a stronger position to secure funds and permits, and to gain knowledge about the location of mineral deposits. Those who lacked the capital to open their own mines joined in as members of the labor force. Villagers whose contracted land was occupied by mining activities, waste, and tailing ponds received some compensation.

Overall, villagers' gains were limited (certainly as compared to those of the state-owned mine and its workers), since they were only allowed to mine where mineral quality was lower, and they were obliged to sell minerals to the state-owned mine for a reduced price. But, as compared to the livelihood strategies they had embraced thus far—farming and limited employment—mining offered opportunities for much better income and without requiring migration. As a result of increased mining activities, damage to the "natural capital" (the environment) soon became apparent. Particularly in areas of the village situated downstream from extraction and processing, such as the subvillage of Fengcun, water pollution became a severe problem. Villagers recalled that at that time the water in the local stream turned black and smelly; they would develop itchy skin rashes if they came into contact with it; and shrimp and fish died.

Villager-run mines increased considerably in number during phase four (early 1990s to 2007), when the state-owned mine was privatized. At this time, most villagers either opened their own mines or formed small cooperatives with neighbors and friends to pool resources (see figure 4.3). Villagers' income from mining increased substantially, and many remarked that all new houses in Qiancun were built with income from mining. More opportunities opened up as side effects of the mining boom. Locals could become businessmen or provide services such as selling vegetables in the market, opening restaurants, and operating minibuses for transport, as Qiancun village experienced an influx of migrants. Gains, however, were unevenly distributed. Those who had already established mining operations by the early 2000s were well positioned to benefit from another peak in the price of lead between 2004 and 2007. By contrast, however, those households that had not yet invested in mining by 2006 were unable to

**Figure 4.3**
Small temple to the Earth God by the entrance to a private mine shaft, 2010.

secure sufficient capital to mine successfully before the new closure policy was issued in 2007. Members of such households lost (the albeit limited) capital they had invested and earned nothing in return.

Loss of natural capital, including loss of farmland and water, and pollution of water, soil, and crops, was substantial. This severely compromised villagers' ability to make a living from farming or even achieve self-subsistence, and forced them to buy food from the market. While conflicts over resources between villagers and outsiders, particularly the former state-owned mine, remained an issue as they had been previously, tensions between villagers were also on the rise, as the opportunities for income rose. In one case, tensions escalated to such an extent when two mineshafts converged that one party involved the local mafia, a villager was killed in a fight, and others were jailed.

The year 2007 signaled the start of phase five, characterized by heavily restricted mining by villagers and limited livelihood options. At this time, the price of lead began to fall, and the Xiangxi district government issued the "2007 Lead and Zinc Industry Pollution Treatment Plan," which demanded the closure of all illegal mines and the adoption of safety regulations for all mining-related activities (interview with township official by Jixia Lu, 2011). These regulatory efforts were strengthened in 2008, following a serious accident in an illegal mine in Shanxi province, in which a slagheap collapsed, killing 254 people and causing other serious injuries (Xinhuanet 2008). Given that the Chinese government still encourages lead mining (Li 2010; Ministry of Land and Resources 2012), this tightening of regulation was not intended to curb lead extraction and processing as a sector. Indeed, new mines were opened and approved by the government in the township neighboring Qiancun in 2012. In practice, the policy most adversely affected those with limited funds and connections to secure permits or to avoid a crackdown. In general, the only mines that continued in operation or resumed operation were contracted by wealthy outsiders. Most villagers by contrast were forced to cease mining operations (see figures 4.4, 4.5, and 4.6). In response, villagers attempted to diversify their livelihood strategies, by seeking work elsewhere or returning to farming (see figure 4.7). A survey by Ran Shenhong (a member of the multidisciplinary team of which I was also part, see below) showed that the tightening of regulations led to the almost total substitution of income from mining to income from migration-related work (Ran 2012). By contrast, mining had literally

**Figure 4.4**
A blocked mine shaft in the wake of the mining ban, 2010.

**Figure 4.5**
The resilience of private mining: minerals deposited outside the nearby shaft to be sorted by quality, 2010.

**Figure 4.6**
Local women select recently extracted minerals, 2010.

**Figure 4.7**
Locals maintaining alternative livelihoods: tobacco growing, 2010.

undermined farming, and most of these damages could not be recovered even after the crackdown. By 2011, mining had percolated through more than thirteen hectares of Qiancun land, with almost two hundred abandoned mine shafts, eleven smelting factories, and three tailing ponds. In recent years, agriculture in Qiancun village has been facing a food safety crisis, though the most recent tests suggest that food safety impacts are complex and uneven (N. Chen 2013; Holdaway and Husain 2014).

The environmental costs of mining are severe (Gao, Shen, and Wang 2012). Lead mining in particular is well known for its environmental and public health threats (Li, Ji, Yang, and Li 2007; Li, Ma, Kujip, Yuan, and Huang 2014; WHO 1995; Zhang et al. 2012). The low efficiency of the lead-zinc mining industry severely impacts the environment by affecting crop yields, use and degradation of water resources, and soil subsidence, all of which undermine agricultural livelihoods (Li, Ji, Yang, and Li 2007). In turn, this generates serious risks to human health on a national scale (Li, Ma, Kujip, Yuan, and Huang 2014; Zhang et al. 2012). Even at very low levels, exposure to heavy metals such as lead, cadmium, mercury and the metalloid arsenic is highly toxic and damaging to the central nervous and cardiovascular systems (Li 2012; Zhang 2011; Zhang et al. 2012). For instance, exposure to lead through respiratory and gastrointestinal systems can cause impairments in brain function, kidney damage, infertility, miscarriage, increased blood pressure, hypertension, and lower IQ levels among children (WHO 1995; Silbergeld 1996; Tong et al. 1998). While some of this pollution is well documented, distant, inaccessible small mines and smelters may be associated with serious, undocumented pollution (Zhang et al. 2012, 2270). Qiancun is one such case. Artisanal small mines (ASMs) are widespread in China, and Qiancun is typical of the landscape of ASMs by including several relatively small and low-budget mines. The ways in which mining is situated in the locality—spreading benefits unevenly across social groups, causing conflicts, and causing a range of effects on livelihood, soil, agriculture, environment, and health—is also typical of mining areas in China, though the extent and type of effects will of course differ between localities (Zhang et al. 2012).

The severity of pollution in Qiancun has been documented since 2000 by a team of medical geographers from the Institute of Geographic Sciences and Natural Resources Research (IGSNRR, Chinese Academy of Sciences). During an early fieldtrip in 2000, researchers were told that water used for drinking, washing vegetables, clothes, and household implements was pol-luted, and that some water sources had become dry, forcing villagers to seek alternative water sources. Villagers also told researchers that fish and ducks had died, and they raised concerns about water pollution with county authorities. The research team carried out pilot studies on levels of heavy metals, including lead, in the soil, water, local crops (including staple foods such as rice and vegetables such as peppers), and in the hair and blood of local residents. Follow-up studies since 2009 have been supported by FORHEAD grants (see below). While data collected is based on limited sampling, and therefore cannot provide a complete picture, it nevertheless offers some insights on the levels of damage and exposure.

Research carried out by IGSNRR highlighted several pathways of expo-sure to heavy metals pollution. It showed that the most significant path of exposure is from rice, while water and vegetables play a much less promi-nent role (see, for instance, Zhang 2011). It suggested that lead content in the surface soil in paddy fields seriously exceeded national safety levels (Li, Wang, Yang, and Li 2005; Zhang 2011), though this may not be used to extrapolate levels of lead in the crops in any straightforward way (N. Chen 2013). National standards of environmental quality for the soil issued in 1995 (Ministry of Environmental Protection 1995) recommend lead con-tent below 250 µg/g,[3] but in Qiancun lead content in the majority of sites tested exceeded this figure fivefold, and in one case it almost reached 2500 µg/g, with cadmium (Cd), mercury (Hg) and arsenic (As) also in excess (Li 2012). Soil pollution, however, varies in different areas of the village depending on proximity to mining, slag heaps, tailing ponds, and water flows (Holdaway and Husain 2014).

Lead content in rice in some fields also exceeded the national food safety standard, with lead content up to 2.2 µg/g, well above the recommended maximum of 0.2 µg/g. Cadmium, mercury, and arsenic were also in excess (Y. Li 2012). Effects on crops may also be uneven, depending on the type of rice grown, the water sources used, and the location and soil composition of the fields in terms of soil pH, quality, and moisture (Holdaway and Husain 2014). Indeed, the most recent tests showed that levels of heavy metals might have decreased since the closing of local mines (N. Chen 2013). However, some of the other effects of mining on the environment—such as the lowering of groundwater levels and consequent drying out of former paddy fields—are irreversible.

Mining affected groundwater levels and quality by reducing the available amount of drinking water and polluting it. Most villagers believed that the quality of surface water, such as in the local stream and ponds, declined as a result of mining. Tests on surface water in Qiancun revealed the majority of samples contained lead in excess of China's water category V (0.1 mg/l), reaching levels of 0.25 mg/l in some samples (Li 2012).[4] Such high heavy-metal contents exceed the river's capacity to dilute them and gradually make the water source cleaner. Given that 51.2 percent of irrigation water is derived from surface water, this severely affects food safety (Ran 2012), clearly highlighting that pollution of water and crops are complementary problems, which require complementary responses. Levels of pollution have varied over time, and probably peaked during periods of intensive mining and processing activity, particularly before the ban on small-scale mining was enforced in 2007. In particular, the decrease of mining and processing activities since 2007 may have decreased pollution.

Such severe, if uneven, levels of pollution resulted in loss of human capital in terms of the health impacts of heavy metal exposure, from workplace exposure, and from contaminated water and crops.[5] In 1991, the United States' Centers for Disease Control and Prevention lowered the recommended blood lead action level for exposure in adults to 15 µg/dL and to 10 µg/dL in children (on whom the effects of lead exposure is more severe), but have since stressed that this is not a toxicological threshold and, in 2012, further lowered the blood lead action level for children to 5 µg/dL (Centers for Disease Control and Prevention 2012). In Qiancun, however, most villagers tested were found to be in excess of maximum recommended levels, with some reaching up to 60 µg/dL. Blood levels of cadmium, mercury, and arsenic were also in excess (Li 2012). Eighty percent of those working in mines suffered from lead poisoning (Li et al. 2005). While not necessarily representative of the entire village, these findings provide a sense of the severity of the environmental health costs of mining.

This short sketch outlines the diachronic shifts in villagers' gains and losses from mining, and how their attitudes and responses to mining changed accordingly across the socioeconomic spectrum. Against such a backdrop of severe environmental and health damage, the local population has made several efforts to draw more direct benefits from mining activities (briefly described earlier) but also to protect the local environment. While according to IGSNRR research, in Qiancun water is secondary to rice as a source of exposure (Li 2012), water was the foremost concern for the local population. Living in the most downstream of four subvillages in Qiancun, Fengcun villagers were particularly worried about water pollution and its effects on health. They often remarked that "this water makes you sick" or that "the water is poisoned." They argued that water pollution was particularly severe when mining was in full swing (until 2007) and local processing plants were in operation. They pointed to several tailing ponds and suggested that these affected fields beneath and adjacent to them, particularly when they collapsed. However, they were more concerned about the effects on local water sources.

How might we make sense of villagers' overwhelming focus on water pollution, despite the fact that it poses a lesser threat than contaminated rice? What are the origins and effects of their focus on water? As they struggle to understand the local occurrence of disease and pollution, especially where little scientific evidence is available, local communities may become fixated on certain aspects of the problem and ignore others (Holdaway and Wang 2013). My fieldwork in Qiancun suggests that villagers' focus on water is likely to occur for three reasons. First, water pollution was most obvious for all to see, smell, and taste. Even during my research in 2010, when hardly any mining and no processing took place, several villagers were confident that, if given two bottles of water—one from a relatively clean source and one drawn from closer to former mining and processing areas—they could taste the difference. One remarked that many young villagers who return after a period away find the local water tastes "strange." Contamination of crops, by contrast, is much less perceivable. Had villagers been aware that rice was the foremost path of exposure, they would doubtlessly have been much more concerned with it than they currently are. However, as I will explain in the final part of the chapter, disclosing more information about these paths of exposure is not something that the project team has been in a position to do, nor something that the local government has deemed feasible so far, however fair it may be in principle.

Second, acknowledging more fully the extent to which the local environment has been affected by small mines, many of which were opened by villagers, would directly implicate villagers. By contrast, highlighting the fact that water was polluted by processing activities (which were not owned or run by locals) deflected the blame toward outsiders. This is a well-documented pattern whereby the focus of complaints is on pollution created by outsiders rather than by locals (Cheng 2013; Deng and Yang 2013). The attribution of responsibility for the impact of pollution on health is a crucial ingredient in how problems are framed and conflicts negotiated between villagers, polluters, and local governments (Holdaway and Wang 2013).

Third, water pollution seems to present more hope for redress than the prospect of halting mining altogether, particularly when so many locals also earn a living from it (see also Lu and Lora-Wainwright 2014; Ran 2012). Alternative sources of water could be sought, for instance (though as I describe below, this too faced several obstacles). In sum, villagers' concerns are shaped by an overlapping set of factors, including the relative perceivability of pollution, their sense of which of their demands are most likely to be addressed effectively, and finally their sense that drinking water may be provided without compromising their reliance on mining as a means of livelihood.

These attitudes and responses, and particularly the focus on water, are a clear case of resigned activism. They embody forms of engagement with local pollution that are neither fully resigned nor fully opposed to it. They showcase the mutually shaping dynamics between the ways in which pollution is experienced and understood and the pathways of action embraced in response. As we shall see more in detail below, villagers adjusted demands and expectations to what they thought might be the best or only option. They continued to complain, but they have learned that their effects in securing a healthier environment can only be limited. In turn, this has affected the ways in which they value mining, the local environment, and their health (see Lu and Lora-Wainwright 2014).

Given villagers' focus on water as a path of exposure, the rest of the chapter will focus on water, through three case studies: (1) petitioning by an educated local; (2) villagers' various efforts to access drinking water; and (3) the involvement and dynamics of the research team, negotiations with the county government, and debates surrounding appropriate intervention. These three cases will shed some light on the entanglements among local social relations, village politics, and the role of research and politics in shaping complex decisions over intervention.

## "Madness" and the Politics of Resigned Activism

One of my biggest regrets in Fengcun is that I never met Li Fang. Early on in my first field visit in 2010, I heard several people talk about a local "educated man," who petitioned repeatedly about local pollution. Unfortunately, he had died of cancer in 2008, at age eighty-five. On August 27, 2012, I spoke at length with his son Li Ping about his father's endeavors. Li Fang's personal life had been all but straightforward. Li Fang was the son of two teachers, and he too had started his working life as a teacher, before becoming the head of the county's food bureau (*liangshi ju*). His first wife (Li Ping's mother) was a teacher in Jishou city. Because of Li Fang's "bad class identity" (a label ascribed during the Mao period to those identified as potentially counter-revolutionary, for instance former landlords), in 1960 he and his family were sent down to the countryside to Xinchang village, not far from Fengcun. His wife died of tuberculosis in 1962, when Li Ping was only nine years old and his younger brother was five. Li Ping bitterly reflected that his mother could have been cured, but there was no medical treatment available in the village. After her death, Li Fang's eldest daughter took her five-year-old brother with her to Guiyang (the provincial capital of nearby province of Guizhou), where she looked for work, while her brother went to school. She could not take both brothers, so Li Ping moved with Li Fang to his natal village, Fengcun. Li Fang never learned how to farm, so he operated a small vegetable oil processing establishment, as he had done in Xinchang.

During the Cultural Revolution (1966–76) he married a widow, but she suffered with a heart condition and died shortly after. He later married another widow, who survived him and now lives with her son from her first marriage. Li Fang's "bad class identity" haunted him, and he petitioned to Beijing to be rehabilitated, finally succeeding in 1975. Instead of leading to a more relaxed existence, however, his rehabilitation fuelled Li Fang's attempts to prove his worth to the people and to the state. Deeply critical of local pollution, in 1985 he filed a petition in Beijing, complaining that the provincial environmental protection bureau had tested the village water but not released any results. Dissatisfied with the lack of any outcome, he petitioned again in 1987, when, according to the village head at that time, there were seven petitions from the district and five from Fengcun alone.

Li Fang's efforts concentrated on proving the harmful effects of pollution on the local population, trying to restore a clean environment, and providing clean drinking water. In view of the first aim, he frequently communicated with the village schoolteacher (in his fifties at the time of fieldwork) to collect information on children's performance in school. Teacher Li told us that Li Fang had consulted him on this topic and that there had been research in the 1980s that suggested that exposure to lead impaired children's cognitive functions and that adults could be affected too. One of the local doctors also told us he was consulted by Li Fang to provide records about local disease incidence that he could use in his petitions. In the hope of restoring a healthy environment, Li Fang drafted new plans for the village, proposing strategies to limit pollution and its effects by building all houses close together along the village road, farming fields with good water, and planting fruit trees in the remaining fields. Much to his frustration, none of these plans ever came to pass. To tackle the pollution of drinking water, he proposed drawing it from a reservoir from a subvillage upstream from mining operations. The village head rejected this proposal on the basis that it was unrealistic; it would cost too much money and the pipes had to cross too many fields. According to his son Li Ping, Li Fang continued to demand redress and make proposals for a cleaner village until he died in 2008.

Li Fang's "bad class identity" is a defining feature of his entire life experience. As a direct or indirect consequence, his status in Fengcun was mixed at best. Some stressed that he had been "a good man" who "cared about the community." Others claimed he was "mad" (fengzi), or "mentally ill" (shenjing bing). His son Li Ping sharply accounted for both views. His argument implied that Li Fang's "madness," his incessant efforts, and his "bad class identity" were related in several mutually productive (and destructive) ways.

First, on a most basic level, his "bad class identity" tortured him and caused emotional instability and mental illness:

Locals called him mad; they did not support him. Actually, he was somewhat mentally ill [you yi dian shenjing bing], he was troubled [xin fan], because he was accused on unjust charges [because of his bad class identity]. He was not at peace. He could not accept it in his heart. ... He had no support from the county government because he was sent down to the countryside, his identity was not ideal, so what he said had no value. ... His heart was broken [bengkui]. It destroyed his nerves and brain.

Second, his "bad class identity" made him feel isolated, pushing him to constantly try to prove himself, and this, too, eventually broke his heart and drove him toward insanity. Indeed, being rehabilitated seemed to make Li Fang even more frantic in his efforts to contribute to the welfare of the state and the people. Third, and conversely, Li Ping suggested that only a "madman" could come up with proposals that were so far from reality. Likewise, only a "madman" would insist on these proposals when he clearly lacked the power, authority, connections, support, and funds to see them through.

With everyone's support, things can be done, but if nobody supports you, no matter how much you think, it's all for nothing. ... Others in the village are not like him; if they have a good life they are happy, but not him. He was concerned for people; he worried in vain. If people like him were more common, this village would be easy to sort out. If everyone cared for the people, not for themselves, it would be easy. ... You tell me: you have no power and no money. You ruin your health. Nobody supports you. Why would you go [to petition]? He invested his own money to petition. It's meaningless, and people said he was mad. He proposed to get water from a reservoir, so people called him mad. Not all opposed him, there were differences of course. But the water had to cross so many fields. Without money, it's impossible. Unless you have hundreds of thousands of yuan, and do it yourself, people will not trust you. First, you have no money, and, second, you have no authority. *What are you other than mad?*

Li Ping was convinced that his father's unrealistic aims and relentless, single-handed efforts were not only a sign of his mental instability and emotional turmoil, but that they also led to his demise:

I told him at his age he should not worry: "Nobody is worried. Why would you worry all on your own? Concern yourself with things that should concern you!" He wouldn't listen. He would say: "I am a state official [we are not aware of him holding an official position in the village, but perhaps he referred to his previous position in the food bureau], I rely on the state; I want to help the state do good things." Doing good things is good, but if you have no money and no authority you cannot do them. I told him not to worry, he didn't listen. ... Struggling alone has no results. Many things are not so simple. With no money and no power, it's no use speaking empty words. ... He was at home writing every day; he wrote till bedtime. I told him to stop and go for a walk: "It is good for your health." But he just sat and wrote. ... His brain never rested, not even when he slept. If he had rested he could have lived longer, instead he died of cancer.

Li Fang's "madness" may be understood in several ways. I am not in a position to establish whether Li Fang was indeed mentally ill, nor should this be the aim of this discussion. Rather I am interested in how this label came about and with what consequences. On one level, it is a product of his isolation—both through his class identity and his failed attempts to gain redress for pollution. On this front, he was unlike members of local elites who play a key role in some of the cancer villages studied by Chen and his team (Chen et al. 2013). In Qiancun, as I outlined earlier, members of the local elite were probably too closely invested in mining to play a prominent role in opposing it. By constrast, emotionally damaged and frustrated by his inability to make a positive contribution, Li Fang embraced activism to restore his sense of civic self and oppose injustice. His refusal to succumb to the rational conclusion (as outlined eloquently by his son) that his agency was severely limited by lack of support and lack of funds was a further element contributing to his being labeled "mad."

On a Foucauldian note, his "madness" is a symptom of his proposals being extremely different from the status quo at the time. It is striking that Li Fang's planning proposals are in fact rather similar to recent state policies promoting stricter zoning to separate industrial, residential, and ecological areas. They also resemble some of the principles behind the "new socialist countryside" program launched in 2006, which include more clustered residential areas close to the main roads in order to ease transport and communication, and centralize the installation of electricity, water sanitation systems, and piped drinking water (Ahlers and Schubert 2009, 2013). In the current policy climate, his proposals might not have been as quickly dismissed as "mad," though the lack of funding to implement them would still have undermined his efforts. When he made these plans, however, they were out of step with the political economic climate. In this sense, his "madness" has more to do with his refusal to conform and his troubled political past than with his ideas per se. This is not to rule out that he may have in fact suffered from mental health problems, however loosely defined—and given the extreme emotional and psychological strains he suffered, it would hardly be surprising (see Kleinman 1986)—but to put his position in its social and historical context.

Li Fang's experience is significant for an understanding of China's rural environmentalism. His social and political isolation are largely due to his own troubled class identity, but his difficult position is similar to that of community-based activists more broadly (including beyond China), who are often isolated, stigmatized, and labeled as "oddballs" (see, for instance,

Brown and Mikkelsen 1997). From a systemic perspective, Li Fang's efforts are barely visible, materializing only at certain moments when he sought redress beyond the confines of the locality. However, his persistent endeavors undoubtedly deserve to be recognized as instances of environmental activism. While his attempts to gain redress largely failed, they had a definite impact on current attitudes toward pollution. Although he may not have succeeded in securing safe drinking water, Li Fang was probably instrumental in alerting locals to pollution. For instance, the comparatively high recurrence of tests is likely to be a direct consequence of Li Fang's efforts over the years to compile evidence and demand redress. Even though test results were typically not released, their frequency strengthened villagers' sense of the severity of local pollution.

On the one hand, Li Fang's failure to gain redress, or even the release of test results, gradually reinforced locals' sense that their environmental plight would be ignored and any efforts to address it would be futile. His experiences fostered a love-hate relationship with outsiders, whether they were representatives of higher levels of government or researchers: they were seen as potentially helpful, but experience to date suggested to locals that little positive change would result from these encounters. This fueled locals' frustration at the extent to which their plight was seemingly ignored, and therefore at their own powerlessness. The fondness with which Li Fang is remembered in the present is likely to be a consequence of locals' desperation over continued lack of attention and lack of resolution, worse still after mines closed in 2007 and income opportunities became more limited.

On the other hand, Li Fang's resilience served as an example that other locals emulated. Most males in Qiancun share the same surname, creating fairly strong kinship solidarity and social cohesion. The presence of a well-educated man who was willing and able to compile petitions helped to channel villagers' unity toward demands for a cleaner environment rather than only demands for compensation (as was the case in Baocun) or for the permission to mine. It has made them relatively outspoken and confident about the harm of mining. In 2011, for instance, a group of locals composed a letter to the local government in which they collated complaints about pollution, what they considered evidence of higher cancer rates linked to it (whether this is indeed the case is a different question entirely), and a list of compensation amounts owed. In a classic case of resigned activism, villagers continued to make such demands, even when they were skeptical about the likelihood of success. Their ongoing engagements to secure safe drinking water offer perhaps the clearest example of relatively invisible (and ineffective) activism, failed attempts at local coordination, and deep-seated concern about pollution.

## Water Projects: Pollution, Trust, and the Politics of Responsibility

Villagers' views of water pollution and their efforts to access scientific evidence and secure safe drinking water provide a good prism through which to view and understand local social relations, political economy, conflicting strategies, and tensions with local officials. Locals' preoccupation with water was characterized by a deep sense of loss. Elderly villagers nostalgically recalled that Fengcun was famous locally for its fresh and abundant natural spring. Its position near the center of the village and the solid structure—with large stone slabs on four sides of a pool measuring almost a cubic meter and a stone channel leading the overflow to the local stream—suggested this had been a key source of water for a long time and perhaps even the reason why the village had been built there. When I visited Fengcun in 2010 and 2012, locals still washed clothes and vegetables at this spring, and occasionally collected water for drinking, but they widely argued this spring was no longer safe. The water had been tested several times over the years, though results were rarely communicated to villagers. On some instances, the environmental protection bureau declared the water was "not drinkable," but most of the time villagers were not informed. They took this reluctance to disclose results as evidence that the water must be polluted, otherwise, they quipped: "Why wouldn't they tell us?"

According to former village head Li Qun, the first water test in Fengcun was carried out in 1972 by the government's mineral search unit. At that time, the state-owned mine dug an eleven-meter-deep well on the hillside behind Fengcun. The water was plentiful, but it was found to contain lead in excess of safe standards. The mine concluded they would have to dig at least fifty meters deep to reach safe water, but instead of doing so they abandoned it due to lack of funds, and the well has since been blocked up. More tests were undertaken in the late 1980s by the provincial and district environmental protection bureau, most likely in response to Li Fang's petitions. The village head proposed spending 60,000 yuan to dig a well, but other officials were against it. Some (including Li Fang) proposed drawing water from a reservoir located upstream from mining operations, but the village head opposed it on the grounds that the reservoir was too far, it would cost between 200,000 and 300,000 yuan to connect the village, and the pipes would have to cross too many fields (thus the system would be too costly to maintain). As a consequence, no solution was implemented. Further tests carried out in the 1990s by the Hunan Province Labor and Health Institute of Occupational Disease Prevention again confirmed some local water sources were unsafe for drinking (Li et al. 2005; Zhang 2011). Yet clashing views on what was feasible and affordable stood in the way of providing clean drinking water.

Such recurrent tests, combined with frequent refusal to communicate results to the public and failure to provide safe drinking water, frustrated villagers and made them feel helpless and ignored. Disillusioned with local cadres, and tired of waiting for an agreement that proved so elusive, villagers have long attempted to take the matter into their own hands. Since the 1990s self-organized groups of households repeatedly bought pipes and searched for water sources. Groups of as many as thirty households paid 1,000 yuan each and contributed labor time (usually between one and two months). Often, however, these water sources became dry shortly after they were accessed, and villagers had to start the process all over again. Most worryingly, such water sources are usually not tested. Villagers assume that water from "up the mountain" is safe, but heavy metal content may still be high.

To tackle these challenges, two state-funded drinking water projects were put forward to cover the entire administrative village of Qiancun. The first was implemented in 2000. It was voluntary and it involved participating households paying one hundred yuan each, with additional funds provided by the state. Water was connected for a short time, but when pipes broke there was no system or funds in place to fix them, and therefore the project failed. The latest drinking water project was implemented in 2009, and it too proved to be a failure. Every household had a pipe installed, but the water was never connected. An initiative intended to provide a badly needed public service became yet more evidence of an uncaring state and of outright corruption. Rumors among villagers had it that the project had been allocated 400,000 yuan by the county. During my fieldwork in 2010, the party secretary claimed he had not yet received any funds from the county and had found them independently (most likely from the township). Old party member and former village official Li Xi was among a group of villagers who went to the county's water bureau to seek clarification and redress. But bureau officials allegedly told them that they had already handed over funding for this project, and now it was in the village's hands. Subsequent conversations by other team members with county leaders suggested that any efforts to address access to safe drinking water in Qiancun to date had not been funded by the county's program for providing such access, leaving the question of where funding came from unanswered. Such confusion and disappointing lack of results led many villagers to conclude bitterly: "The officials ate the money." Some claimed that the village secretary was in fact drilling for minerals, not water. The failure of this initiative was detrimental, not only because it still left villagers without safe drinking water, but also because it further eroded villagers' already abysmally low opinion of the village party secretary and their trust that the local government more broadly would provide for even their most basic needs. It reinforced their sense that complaining has no results, that the local government does not care for them, and that even the presence of state funding cannot help them.

The only way, villagers were left to conclude, was for them to find a solution. Villagers, however, were deeply resentful of being held responsible for finding clean water. Granted, they too took part in mining, but they argued the burden of responsibility for water pollution was predominantly on the former state mine and on processing plants. The collapse of a tailing pond in 2012 and the reopening of some processing plants also reinvigorated demands for clean water. Indeed, in 2012, Li Xi collected more than thirty signatures to demand that the contractor of the former county-run processing plant stop polluting the local water and provide safe drinking water. At his request, the county Environmental Protection Bureau (EPB) tested the water and openly declared it to contain lead in excess of safe levels. In response, the plant contractor provided two thousand kilograms of pipes to the village, and the villagers were told to find water themselves (see figure 4.8). The villagers were angered by this response. Former village head Li Qun complained: "They told us to find a solution by ourselves [*ziji xiang banfa*]! [If] you pollute our well, why should I find a solution? This problem is caused directly by a county industry, so the county should clean up!" When I asked what he would propose, he said: "If I was a leader, I would

**Figure 4.8**
Attempts to gather drinking water in Fengcun, 2012.

find a solution. I would go to the county. Whoever polluted should clean up; the state policy is very clear. There are so many years of pollution left here, and until now it is still unresolved."

Employing a widespread strategy that Kevin O'Brien and Li Lianjiang famously termed "rightful resistance" (2006), Li Qun's reference to "policy" was intended to lend legitimacy to his complaints and his proposal. His disparaging comments about the county echoed similar widespread feelings of frustration among villagers. Being burdened with having to "find a solution by ourselves" once again exacerbated villagers' mistrust of officials. It also fed their resignation to the absence of a reliable long-term solution. Their own attempts to secure safe drinking water were a manifestation of the lack of coordination within the village. In 2012, pipes were scattered everywhere across Fengcun, in an effort to connect them to water sources further up the hillside presumed (but not tested) to be safe. The ubiquitous presence of pipes is a sign of how concerned they were about "poisoned water." It is also a visible symbol of their inability to organize in any sustainable way. The failure of their significant efforts over the past two decades to secure clean water has eroded villagers' trust in the possibility that even organized demands and efforts might result in any positive change. Li Fang is a reminder of the futility and potential harm of trying in vain to obtain redress. The inability to benefit directly from mining after 2007 and yet being still exposed to its effects has made villagers feel at once restless and helpless. They persisted in attempting to gain attention and redress, and yet felt that their efforts made little difference. While the closure of most mines since 2007 may be seen as an improvement in environmental enforcement, it seemed to leave existing pollution untouched, as Li Qun pointed out.[6] These closures in turn posed serious challenges for local government budgets, which, deprived of income from mining, are even less in a position to provide public services like drinking water (Tilt 2010).

Several points can be drawn from this short chronological description of villagers' efforts to secure clean water. Reference to challenges posed by cost and feasibility made by those in charge (be it at the village level or higher) may be an effective strategy to justify not tackling water provision, but these challenges are no less real. Indeed, in the case of connecting Fengcun to a reservoir several kilometers away, the lack of sufficient funding to maintain the pipes would make the entire project infeasible in the long term, and result in a misuse of what little existing funds may be available (see also X. Wang 2013). Conversely, connecting to new water sources that are presumed to be clean, but not confirmed to be safe through tests, would also consume funds that may be better used in other ways.

Confusion over how to finance and organize access to clean drinking water in a site where supply is so limited is a real problem, not one that is simply fabricated by local officials to protect their positions. As all these failed efforts testify, the provision of clean drinking water in Qiancun is an extremely complex problem, with no easy solution. Indeed, my aim is not to argue that local officials behaved in a corrupt fashion: I am in no position to establish whether this was the case. What I can conclude based on fieldwork findings, however, is that continued lack of clean water had deep social and political consequences. It strengthened villagers' distrust toward village leaders and the local government. Under these circumstances, even if a technical difficulty or absence of funds is real, the lack of transparency and lack of respect for local cadres rooted in decades of experience understandably leads villagers to assume corruption is at least one of the reasons for the sustained lack of resolution.

These findings on Fengcun villagers' troubled search for drinking water offer some rich material for understanding the environmental plight from a local perspective, and for examining the complexity of local social relations, the relative cohesiveness of the local community, and the roles of local officials and outside interventions. They show that the types of action and activism embraced in the search for clean water are powerfully shaped by evolving parameters for what is seen to be possible, who is held responsible for pollution, and what previous experiences may have taught villagers. In the process, activism and resignation intermingle as locals struggle with limited evidence and limited funds to support their claims. In some ways, their experiences are mirrored by the evolution of the multidisciplinary project of which I was part. I turn to this next, in order to examine the challenges encountered, and the contexts and negotiations affecting pathways of experimental intervention embraced to date.

## Multidisciplinary Research and the Challenges of Intervention

My work in Qiancun was part of a research team brought together by the Forum on Health, Environment, and Development (FORHEAD), a capacity-building and knowledge-sharing platform supported by the Social Science Research Council's China Environment and Health Initiative (CEHI) and the Institute of Geographic Sciences and Natural Resources Research (IGSNRR) at the Chinese Academy of Sciences.[7] As it was no longer feasible for me to be actively involved in the project, and least of all in the field-work, after 2012, I am not in a position to comment at length about developments since then, but neglecting to mention them would present a partial view of the project's activities, aims, and outcomes. Much of my account on developments since then therefore draws heavily upon correspondence with, and invaluable updates from, other project members, particularly Wang Wuyi and Jennifer Holdaway.

Medical geographers at IGSNRR began to research the impacts of mining on health in Qiancun in 2000 (see section on environmental health impacts earlier in this chapter). Soil, crops, and water samples, as well as hair and blood samples from local residents, demonstrated serious pollution levels by multiple heavy metals (Y. Li 2012). However, given several obstacles, the local government and the local population seemed to take limited and ineffective action to tackle these threats. In order to gain a more comprehensive understanding of the environmental health impacts of mining, villagers' risk perceptions, the history of development in the area, and its broader cultural, social, economic, and political contexts, FORHEAD funded research in the area through several small grants. In its initial funding phase, FORHEAD supported further work by IGSNRR researchers. This was followed by grants that provided for the inclusion of social scientists alongside natural scientists. This multidisciplinary team included experts in environmental health risk assessment (Wang Wuyi, Yang Linsheng, and Li Yonghua, Chinese Academy of Sciences), risk perception (Zheng Rui, Chinese Academy of Sciences), public health (Fang Jing, Kunming Medical University), land use and ecosystems services (Ran Shenghong, Chinese Academy of Sciences), and rural development (Lu Jixia, China Agricultural University), and CEHI's director (Jennifer Holdaway, SSRC), who coordinated research design and facilitated multidisciplinary collaborations. I contributed an anthropological angle to the team, with additional support from the British Academy. More recently, other researchers joined, particularly soil remediation expert Chen Nengchang (Guangdong Institute of Eco-environmental and Soil Sciences) and Su Shipeng (Fujian University of Agriculture and Forestry), focusing on policy. Throughout each subsequent project, the team worked with the vital support of the local Center for Disease Control and Prevention (CDC) in Fenghuang County. Medical geographers were the first to establish Qiancun as a research site and create a collaborative relationship with the county government. Therefore, they acted as extremely helpful contact points for research in Qiancun by other team members, helping to mediate arrangements with the CDC, who in turn assisted with fieldwork practicalities, including accommodations in the village. IGSNRR researchers also continued to play key roles in the project through their own research and engagement with local authorities.

Members of the multidisciplinary team made their first group visit to Fenghuang in 2009. We discussed some of the CDC's main local initiatives (such as HIV/AIDS education programs), the main public health problems in the county, the challenges they faced, and ways in which our research might assist them. Over the course of 2009 and 2010, team members corresponded regularly to revise several drafts of a questionnaire, which included questions on family economy, land use, health, environment, diet, and risk perception. In August 2010, I traveled to Fenghuang with team members focusing on risk management, psychology, and rural sociology. In the course of a week, we visited the township government, where the Qiancun village secretary, the village accountant, and several other village officials, including the Fengcun village head, offered an introduction to the locality. They stressed that the area was poor and that villagers relied heavily on mining for their livelihood but they were also wary of pollution. They highlighted poor health among locals and the shortage of safe drinking water, but quickly explained that they lacked the capacity to solve any of these problems. On this basis, they requested our help in gaining attention from higher government levels, so they could secure funding to tackle these challenges effectively.

During one visit to Qiancun, we held a focus group discussion with former workers at the state-owned mine, who faced a particularly difficult situation. Since the mine had been contracted to a private company, many had lost their jobs. Worse still, since the mine had ceased operations in 2007, recently retired workers had no pensions. While villagers could fall back on farming, formal mine workers were officially classified as "workers" rather than "farmers," and therefore had no entitlement to an allocation of village farmland. On another visit to Qiancun, the village doctor helped to gather half a dozen neighbors willing to respond to the pilot questionnaire (see figure 4.9). Upon returning to the county seat, we discussed which questions might need to be trimmed or rephrased and debated how the project might move ahead.

After a week, other team members left Fenghuang, and I settled in Qiancun with an extremely capable research assistant, Lu Jingfang. We lived with the local doctor's family, in the subvillage of Zhaicun, sharing a room with their twelve-year-old daughter, Mei (see figure 4.10). It was a particularly hot and dry summer, and most of their vegetables had not survived it. As a consequence, most of our diet seemed to consist of green peppers and salt-preserved green beans. Eggs were regularly available, thanks to their overactive hens and rooster, which kept us awake most of the night by chasing each other around the courtyard just outside our bedroom window. Somewhat desperate for relief from the heat offered by the single fan in the room, Mei, Jingfang, and I sometimes found ourselves all sleeping in the same bed. During our stay in the village, we did participant observation and at least two interviews per day. At her father's request, Mei often accompanied us during interviews to ease our search for willing respondents. At first, we interviewed neighboring families, but we soon decided

**Figure 4.9**
Pilot interviews in Qiancun, 2010.

**Figure 4.10**
Everyday life in Qiancun, washing clothes at the local well, 2010.

to focus on the subvillage of Fengcun, which was most severely affected by pollution and where residents were most outspoken about its effects. This required a half-hour walk from our house and made us reluctant to return home for lunch, in the scorching heat. When we were lucky, we were offered lunch by our morning interlocutors and sheltered in their home until we could move to our next interview. In the absence of such hospitality, we retreated to a small local convenience shop to have a snack and a rest, or we faced the walk in the sun, with the protection of our small umbrellas, and then set out again in the early afternoon. Key informants accompanied us on long walks up the surrounding hill and through the mining area, which provided us with opportunities to gather a better sense of the local landscape and discuss informally the effects of mining on the villagers' livelihood.

After this first fieldtrip, I reported to the rest of the team on some of my most significant findings. I compiled a timeline with some of the main local events and provided names of key informants to follow up with and some ideas on how to proceed with the next research visit. I highlighted that villagers' main concern was water contamination and the shortage of safe drinking water, something that IGSNRR researchers already knew (see above). I also reported that villagers seemed extremely worried about the health effects of pollution and were very vocal in demanding intervention and asking for our assistance in their efforts. In the coming months, team members continued to refine the questionnaire and strived to find a suitable time for the next field visit that would fit with everyone's schedules. As this proved difficult, some members of the team visited Qiancun in late 2010, others in October 2011, and others in May 2012 to administer the questionnaire and conduct some in-depth case studies. I returned to Qiancun, again focusing my attention on Fengcun, in August and September 2012. This visit yielded some valuable data, but villagers were increasingly frustrated that their "water problem," as they put it, remained unresolved. This made some people reluctant to engage with me, as the opening of this chapter illustrates. They felt that I, like the township and county officials who sometimes visited the village, was incapable or unwilling to offer help.

One of the project's aims was to work closely with the county authorities to better understand the complex situation at hand and to identify potential interventions that they could feasibly implement. To this end, in

November 2012 FORHEAD invited to Beijing a small delegation from Feng-huang County—which included the head and deputy head of the CDC, the head of the Environmental Protection Bureau (EPB), and the Qiancun village doctor. The research team prepared a joint presentation on our findings, and we set aside ample time for discussion with the delegation. I was particularly excited by this prospect: it was a chance to have a genuine dialogue among disciplines and with members of the county government and to reflect on what we, as researchers, could practically do to help. Members of the team took turns to introduce their part of the project. The presentations summarized the local history, land use, and livelihood changes, detailing some of the damage incurred and some of the environmental impacts, health effects, and public health challenges (see above). Risk perceptions and responses were covered by psychologist Zheng Rui and me, representing anthropology. A written report of our main findings and recommendations based on these presentations was later compiled and sent to Fenghuang's CDC.

The conclusions Zheng and I reached about intervention were very similar. In our joint presentation to the delegation, we made two key points. First, we suggested that drinking water was the most prominent concern for locals, even if it might not be the most significant path of exposure to contamination. I argued that villagers' experiences with seeking drinking water has created a vicious cycle whereby top-down assistance was required to support local efforts and sustain them in the long-term, but the track record of previous water projects had obliterated locals' trust in the capacity and willingness of officials to provide clean water. Because of their lack of trust, any potential intervention efforts may not be supported by locals, and were therefore bound to fail. (Of course, there are other potential reasons for failure, such as the lack of government capacity and possibly even corruption.) Conversely, tackling effectively what villagers felt was the most pressing problem—water—would also help restore levels of trust and ensure the long-term sustainability of new water projects. This would hopefully transform the vicious cycle into a virtuous one.

Second, we explained that the local population was extremely worried about health effects and attributed all manners of symptoms to lead contamination, including some that are not epidemiologically related to it. For this reason, we recommended that better information about paths of exposure to lead contamination and the specific health risks it poses should be made available. This, we argued, would not undermine social stability (pre-dictably, a foremost concern for officials), but rather improve transparency and allow villagers to discern what problems might indeed be related to lead poisoning, and which might not. To the contrary, we added, the reas-surance that at least some of their health concerns were not linked to lead pollution could potentially help to appease some of the ongoing unrest among villagers. My conclusions were led by an anthropologically based sentiment. I was convinced that only by making the local population feel heard and tackling the challenge *they* had signaled to us, could we galva-nize public participation; and only by improving mutual understanding, communication, and transparency about the risks could any other pro-posed interventions be effective in the long-term.

Both of our suggested interventions faced several challenges. Let me examine them in turn. In terms of tackling the problem of water supply, there was broad agreement among both team members and members of the Fenghuang delegation that it was an important problem to fix. However, given how limited water sources are (largely due to mining activities affect-ing them), and given the extensive resources needed to map them, and to install and maintain pipes, our team was not in a position to tackle the problem of water supply. We lacked not only the funding, but also capacity and standing to conduct such a project ourselves. The absence of locally based researchers further hampered these efforts. The best we could do was to raise this issue with the county government. Indeed, Wang Wuyi made the case several times to county leaders to prioritize the village for county-funded water projects, and the leader of the CDC, Teng, supported this suggestion. Their efforts resulted in promises by one county leader to priori-tize Qiancun for water projects, but nothing has happened since. The polit-ical sensitivity of heavy metals pollution posed a substantial obstacle, but not the only one. Engaging other local government agencies beyond the CDC, whose involvement would be necessary for a sustained and effec-tive intervention, remained difficult. Limited expertise and equipment at county CDCs and Environmental Protection Bureaus, as is the case in Fen-ghuang, hampers their ability to carry out environmental health assess-ments and interventions (Holdaway 2013; Holdaway and Wang 2013). The comparatively low revenue stream for a poor county like Fenghuang poses further obstacles. In addition, the severity of pollution in other parts of Hunan province may have meant that the team's suggestions to prioritize provision of safe drinking water in this area were overshadowed by even more dire need elsewhere.

Improving the local populations' knowledge of the risks posed by heavy metal pollution and of ways to decrease exposure—efforts characterized by the shorthand "public education"—has been on the agenda throughout FORHEAD-funded research in Fenghuang. The question of what such materials should cover, however, is a complex and evolving one. Soil tests revealed that levels of risk are very uneven across the village and they change over time (N. Chen 2013), therefore the same advice may not necessarily apply across the village. Providing suggestions to villagers about how to reduce risks without support from the local government is also infeasible. As in the case of water provision, any intervention on this front would have to be administered by the local government, particularly the CDC, who would ultimately be held responsible. This is in part due to the political sensitivity of the problem, but also because a relatively knowledgeable and trusted local, like the village doctor, would be best positioned to convey this information to villagers.

Although responsive to our suggestions on public education, and appreciative of our motivations, CDC staff members were also understandably concerned that embarking on risk awareness campaigns in Qiancun would alarm villagers and potentially cause unrest. Reports on "cadmium rice" (rice with a high content of cadmium, which is extremely harmful to health) in 2013 (He 2014) and a general tightening of the political climate since then have made any potential public education campaign even more sensitive. CDC leaders feared that villagers—who already felt tense about the lack of resolution—would worry even more if they were given information about the risks without at the same time being offered redress. Drawing on their experience with health education campaigns about HIV/AIDS, CDC staff suggested first identifying ways to reduce risk that villagers could follow, and only then making more information available about the risks. Ultimately, the choice remains with them about when and how to proceed with public education.

As of 2016, public education materials for farmers based on research so far have been prepared by Chen Nengchang, Wang Wuyi, and Fang Jing, with the input of the local CDC in order to adapt them to the local contexts. These materials will be made available for use in Fenghuang and similar areas. Unfortunately, the role of social scientists in this part of the process has generally been more limited due to the political sensitivity of tackling severely harmful pollution such as from heavy metals. In particular, our inability to live in the village for a sustained period of time and to establish stronger relationships with villagers prevented qualitative social scientists—particularly those like myself who rely on long-term fieldwork and participant observation—from building trust with the local population, which would have laid the foundations for better communication.

While the project team was not in a position to directly provide water or conduct public education, it did have sufficient funding and training to undertake a soil experiment designed to identify ways in which villagers could reduce the health risk from heavy metals in rice. Team member Chen Nengchang also had promising precedents of improvement in other sites. Led by Chen, in 2013 some team members visited Qiancun and planted different varieties of rice seedlings in selected locations, clearly marked with billboard signs "soil remediation pilot field." Samples included four rice varieties with low uptake of heavy metals, to test which worked best. High silica fertilizer was applied in one field to fix metals in the soil and therefore decrease uptake in the crops. Finally, a high-uptake crop was planted in a field affected by the collapse of a tailing pond in 2004. While crops from the other fields were intended for consumption, this crop was intended as a remediation measure to decrease soil contamination. FORHEAD arranged to buy the produce from this field, destroy it, and compensate farmers for their loss.

What could be done by the project team and the local government to tackle the environmental health threats in Qiancun? As explained above, the goals of the project were to understand interactions among mining, development, environment, and health in this context, with a view to providing a more holistic knowledge base to guide responses. Although we hoped that our research could be of direct use in improving the situation in Qiancun, the team did not have the standing or resources to do this alone. Any intervention could only be undertaken with the active participation of the local government in finding resources to address the water supply problem and in permitting public education to reduce risks through food. The team is hopeful that the latter may still be possible, but no action has been taken on the first. Beyond Qiancun, knowledge generated by the project may help local governments faced with similar problems to understand and evaluate the pros and cons of different options for addressing them. It may also be helpful in informing policy at the national level. For example, Chen Nengchang is currently advising on the drafting of the soil pollution law that will be issued in 2017. But for the time being, constraints to improving environmental health in Qiancun remain in place.

Working within these constraints requires coming to terms with the frustration of seeing villagers exposed to several risks, hearing their concerns, and yet being in no position to help them directly. While I and some other team members may have felt an ethical obligation in principle to disclose information to the local population, in reality doing so earlier could have been counterproductive, if the information conveyed was not clear and did not provide helpful guidance to villagers about how to reduce their risk. Done without the consent of the local government, it might also have resulted in ending the project altogether, giving us no further opportunities for engagement and causing serious problems for our partners in the local CDC. Painful as it is to accept, I have come to terms with this ethical compromise as a necessity, in view of potentially more effective, sustainable, and long-term interventions when the opportunity arises. Anyone with the ethical compulsion to "help" needs to face the question of what "helping" means, and the answer may be counterintuitive, not at all straightforward, and require patience, humility, and pragmatism. Learning about the limitations to our project and to my role in it has provided me with an experience somewhat resembling the development of resigned activism. While of course the comparison may seem spurious—my wellbeing is not affected by my ability to intervene or even advise on interventions—my involvement in the project and growing understanding of the locality required that I, too, adapt to the circumstances and come to accept that certain pathways of intervention, however desirable they may seem, might be infeasible.

Public education on its own is not a panacea any more than a ban on mining is. Clear and targeted information on the risks posed by contaminated crops may alert villagers to the wider threats to their health, yet it would do little to address their reliance on mining to secure an income. As CDC leaders are all too aware, informing locals more fully about the risks they face does not decrease those risks. This, of course, should not be a reason to hold back information, but disclosing it may well exacerbate villagers' sense of feeling trapped with a polluted environment, precarious health, and limited livelihood options. Villagers' ambiguous position as both victims and beneficiaries of mining, even as costs and benefits are unevenly distributed, places them in a difficult position. Abysmal levels of trust toward the local government—illustrated briefly by the case of the water projects—fuel this sense of helplessness further. Just as Li Fang's presence and his tireless efforts reinforced locals' awareness of pollution but also unwittingly fed their sense of resignation to it, similarly our project has not been in a position to radically alter their perceived (and real) powerlessness.

Ultimately, I still believe that improving mutual understanding, communication, and transparency about the environmental health risks facing villagers must be the basis for building trust in local authorities and in turn securing locals' support toward any interventions that may be embraced. At the very least, it places villagers in a position to try to help themselves, and to target their efforts more efficiently and effectively in order to decrease the threats they face. Project participants remain hopeful that the planned public education program will take place. But over the long term, the best hope for sustainable results is for this complex bundle of problems—severe environmental health threats, lack of accurate information on these threats, limited livelihood options, and limited trust toward the local government—to be tackled through coordinated, multiple, and multifaceted interventions, if any of them are to be successful at all.

# 5   E-Waste Work: Hierarchies of Value and the Normalization of Pollution in Guiyu

At eight a.m. on a hot June day, Juanjuan and I found ourselves acciden-
tally locked inside her brother's large rented workshop in downtown
Guiyu.[1] Sometimes referred to as "the e-waste dump of the world," Guiyu
has specialized over the past three decades in trading and processing
e-waste, a catch-all term for discarded or waste electrical and electronic
equipment (WEEE) such as computers, TVs, VCRs, DVD players, stereos,
telephones, and mobile phones, as well as "white" electronic goods, such
as washing machines or air conditioning units. Guiyu has gained considerable notoriety since the turn of the millennium, as shocking images of
pollution circulated in the domestic and foreign media in the wake of a
damning report by the BAN (Basel Action Network 2002). The town's abysmal reputation posed significant hurdles to conducting fieldwork, since
locals were keen to avoid any more bad press.[2] Indeed, on my very first
visit the previous year, I was accompanied by my academic host Professor
Li Liping's secretary, and I was not even allowed to step out of the car for
fear that locals would be aggressive toward an obvious outsider. As luck
would have it, one of the first year medical students, eighteen-year-old
Juanjuan, was from a town on the fringes of Guiyu, and she kindly offered
to host me at her family home. Through her family's extensive networks
and their own engagement in e-waste work (a shorthand I use for convenience to include both e-waste processing and trade), Juanjuan quickly
became my key contact, entry point, and lifeline in Guiyu. Given all these
difficulties in gaining access, it now seemed to be quite an achievement to
be stuck there.

Juanjuan's family home was not conveniently placed for the main
e-waste trade route, so her brother rented this workshop to trade plastic (see
figure 5.1). Most of the space was occupied by hundreds of bags of tiny

**Figure 5.1**
Juanjuan's brother's storeroom, 2013.

plastic pellets, piled as tall as the five-meter-high ceiling. There was a small and simple toilet in one corner, and one section was screened off from the rest of the workshop as a living space. A water cooler, a fan, a low table, and some low seats were arranged along the wall, near the double bed where Juanjuan and I had just spent the night. Trade was going through a slow phase, with prices at an all-time low, so the pellets sat there waiting for a more profitable time to change hands and to be eventually refashioned into new plastic goods.

Juanjuan made a string of phone calls, and eventually her cousin—also a trader—reached us on his motorbike from her natal village ten minutes away and unlocked the heavy metal door, finally releasing us onto the street. Juanjuan asked him repeatedly if we could follow him for the day on his trading visits, but he flatly refused, explaining that having a foreigner around would compromise his sales and make his customers suspicious. I could fully understand why, and I found myself wishing, not for the first time during fieldwork in China's polluted areas, that I could have made myself invisible.

Despite everyone's claim that the market was slow, Juanjuan and I were woken up at five a.m. by the sound of heavy traffic. Lorries converged onto the town after a few hours on the road from the previous cog in the e-waste trading machine, most likely Guangzhou and Nanhai. By the time we were released from our impromptu captivity, most large deliveries seemed to have taken place. Still, the road was fairly busy with the occasional lorry and several rickshaws transporting smaller loads. Dismantled plastic casing from PCs, electrical wires, discarded earphones, and old keyboards whizzed past as we made our way across the dusty road. A middle-aged, heavily tanned man (surely a sign he spent much time outdoors) raked bits of old keyboards off the back of a lorry. Down a smaller side street, other workers raked small plastic colorful pellets recently washed so they would dry more efficiently. These relatively lighter jobs were more common among locals, while migrants often took on physically more strenuous or more harmful jobs, just as in Baocun.

As folkloristic as the scene was, I knew I could not take any pictures. Outsiders were eyed with deep suspicion in Guiyu, all the more when they were as visibly foreign as I am. Things were made worse by the latest string of media reports aired on China Central Television about the grim environmental health conditions in the town.[3] Local residents and local officials were on high alert to avoid any more negative publicity, and a white face is typically associated with just that. To mitigate the danger of running into confrontational workshop owners, or, worse still, local police, I took to keeping an umbrella very close to my head (which also helped shelter me from the scorching sun), making my face barely visible but only to those closest to me. On one occasion, I was sure a policeman scrutinized me from the other side of the road, and I quickly walked away with Juanjuan to a nearby market, to hide at the stall where one of her relatives worked. My heart skipped a beat every time a car slowed down, or when the police drove past. I sometimes found myself wishing for the lack of attention I mustered in Qiancun. The expression in my native Venetian dialect, "*a volte massa, a volte massa poco*" (sometimes it's too much, sometimes not enough) frequently came to mind. Not that I was doing anything illegal: I had a research visa and all the required permissions, and I did not have to register with the local police unless I spent more than three days in a row in the area. But the suspense was emotionally tiring.

I was extremely relieved when Juanjuan's friend Lindi whisked us off to his workshop. Lindi was in his early twenties and the third of eight brothers. He lived a stone's throw from Juanjuan's home and given its large size, his family had ten *mu* (6,667 square meters), a sizable amount of farmland. He and his siblings helped with the two harvest seasons, and they sold the produce from one season while keeping the winter crop for family consumption. This barely covered costs, however, so Lindi and his brothers, like most locals, looked for other sources of income. For Lindi, this came in the shape of CD drives, which he bought by weight, dismantled into more than ten parts, which he resold mostly by weight. Because space was limited in his village home, and the location not convenient for trade, he rented a small, simply built bare brick warehouse on the fringes of Guiyu, on the shores of the Lian River, just off the main road and surrounded by farmland. He worked alone most of the time, and relied on help from his family during particularly busy periods. This was not one of them.

Riding on the back of Lindi's motorbike with Juanjuan, I put on my cap and large sunglasses and kept one hand over my mouth. This provided some minimal protection from the dust and the distinctively sweet and eye-watering smell of burning plastic. But for the most part, I was trying to avoid being spotted. Once we reached his workshop, I could finally peel off my makeshift disguise, sit back on a low plastic stool (most probably made from recycled material), and relax. The warehouse was large enough to park his motorized rickshaw and to store some goods (see figure 5.2). As in Juanjuan's brother's workspace, a small makeshift area was carved out from the main space as the living quarters. The room was barely ten meters square, and tightly packed with a double bed surmounted by a mosquito net, a plastic and metal dining table, a small wardrobe, and a small desk on which sat an old CRT television, and a few small parts from the last CD drive he dismantled. Lindi smiled apologetically about the messy state of the room, "I'm sure you can't take this!" I quickly rebutted: "Of course I can. It's very interesting. I'm glad you have time." He sat down and sighed: "It's hard to get by nowadays." When business was thriving, Lindi slept at his workshop, but with stricter controls since the previous year and a drop in the price of metals, few were willing to trade. He showed me receipts for his recent transactions, which were only every fortnight. "You see, there's no work!"

**Figure 5.2**
Lindi's workshop, 2013.

Lindi patiently explained his work while we tucked into some piping hot *changfen* (a sort of rice flour cannelloni filled with vegetables and meat) and sipped cold tea made with herbal remedies to help relieve the body from the summer heat. He bought whole CD drives and first divided them into the plastic case, the metal case, and the reader (figures 5.3 and 5.4). The first two were sold by weight. The reader, if reusable, became an "end-product" (*chengpin*), and was sold whole. Lindi could tell easily if readers were still working, and also where they had been manufactured, based on the quality of the plastic and their serial numbers. When readers no longer functioned, he separated them into smaller parts, including a magnetized flat tape drive, and a smaller core inside the reader. He said taking the reader's core apart was laborious and not very lucrative, so he sold them whole, after extracting two small magnets, which were also sold by weight. He recalled that after the 2011 tsunami in Japan, these magnets were in high demand, but now they were no longer so valuable. "I wish there was another earthquake," he joked. He sold the rest of the reader to a friend who dismantled it and made a small profit from selling microchips and copper parts.

**Figure 5.3**
Lindi's sorted CD cases, 2013.

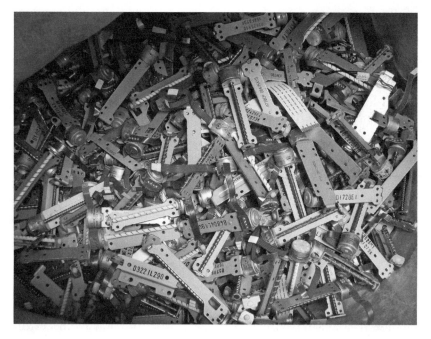

**Figure 5.4**
Lindi's CD parts, ready for trade, 2013.

Different dismantled components of CD drives sat in large plastic bags neatly organized around the warehouse. Some bags contained small strips of magnetic tape, others various plastic parts, others the reader. In one corner, he piled the metal casing and the plastic casing, separating dark from light. I remembered seeing similar piles near Juanjuan's home, and was pretty sure they must have been products of his work (see figure 5.5). He quickly dismissed those piles: "They're just worth a few yuan." He estimated when business was good he could earn thousands of yuan in a day. If he employed workers, he might only make two or three hundred per day, but he never did, relying instead on relatives to help. "End-products" such as working readers, sold for one yuan each, earning him a few hundred yuan per time, and only requiring "light work." Small circuit boards inside CD players sold for twenty yuan a kilo. Some of the capacitors that could be detached from the circuit boards with pliers could become end products (to be refitted as capacitors in a new circuit board), but most were sold by weight at two yuan per kilo and melted to extract their little aluminum content. When he accumulated enough goods, Lindi made a string of calls

**Figure 5.5**
Dismantled CD cases, ready to be traded by weight, 2013.

to double-check the going rate for his materials. He hired a rickshaw for three yuan or a truck for ten yuan, loaded the large recycled bags, and took them to the weighing station, where he was charged three yuan to weigh each load. When he traded with familiar customers, he skipped the weighing process and they agreed on a price based on trust.

Lindi was quick to point out that his business was not among the most lucrative, and certainly not polluting. Metal melting and refining—which involved extracting metals such as gold with *aqua regia,* a mixture of nitric and hydrochloric acid—were the real culprit, and provided the biggest income. Given the severe levels of pollution, few women took these jobs, which were typically the domain of migrant men. "Circuit boards are burnt at very high temperatures and metals flow out. That smoke is very poisonous, but it is a big earner; everyone wants to do it. You may invest hundreds of thousands of yuan, but you will earn it all back in one or two months. I went near one of those ovens once. It's forty or fifty degrees [Celsius] in those rooms. From half a kilo of good circuit boards, you could make almost one hundred yuan; if you could refine five kilos of metals you earn over a million." Would you do that job, I asked? "No, you can lose a few years of your life just doing it for a year. Now these businesses are all hiding in the mountains and working at night. If you want to refine metals, you call them. The minimum quantity is fifty kilos (of circuit boards), you agree on where to meet, you hand over the material for them to process. You can call in the middle of the night and they will do it. But with stricter checks and low value for metals now, even the bigger metal refining workshops are losing money."

This ethnographic vignette raises many questions that will be explored in the course of this chapter: How has e-waste work become so deeply embedded within the local social, political, and economic context? How are costs and benefits distributed? How does this affect the ways in which locals regard e-waste work and its potentially noxious effects? How does e-waste work affect their subjectivity? Understanding the role of e-waste work for local residents requires thinking more critically about all its different components and the vastly different processes it comprises. In turn, this allows us to grasp the conflicting elements that workers, residents, and workshop bosses must face when they decide whether to engage in a particular line of work, whether to live in Guiyu or move out, and whether potential economic benefits are worth sacrificing their health and the health of their families. This case provides another example of the ways in which expectations for what constitutes a healthy life or a clean environment are powerfully shaped by the pervasiveness of pollution. For the most part, residents are actively resigned to living in harmful surroundings and yet endeavor to emancipate themselves from blame for bringing them about.

## Making Sense of E-Waste

Researching citizens' attitudes toward pollution, its health effects, and the actions they may take against pollution and its perpetrators is no easy task anywhere, least of all in China. Doing it in a media hotspot is all the more challenging. In Baocun and Qiancun, pollution, inequalities, and the plight of the poorest are largely invisible, like in so many other places throughout China, and more globally, no doubt. The almost thirty villages that make up Guiyu Town occupy their diametrical opposite on the spectrum of media coverage. The pattern of local development is also different. Whereas in Baocun and Qiancun there was initially one main, state-owned firm, and locals only later became involved in sideline activities, in Guiyu local residents were the main engine of development. For the most part, until 2013, e-waste processing businesses remained relatively small-scale, operating in residential areas, based in people's homes and their courtyards. Despite these differences, a comparison between Baocun and Qiancun, and Guiyu is fruitful. The sense that pollution is inevitable pervades locals' accounts as it did in Baocun and Qiancun. In this chapter, I explore how and why acquiescence toward pollution is engendered in Guiyu and with what effects.

The category of e-waste is as immensely broad and complex as the processes involved in the reuse and recycling of these goods and their components. Some of these processes are relatively clean and involve separating working components (batteries, screens, microchips) from outdated or faulty machines and reusing them in the production of new goods. Others, however, are less salubrious, most notoriously when they involve "baking" circuit boards and using aqua regia and other acids to recover valuable metals such as gold. Without appropriate disposal or waste treatment mechanisms, acids are dumped into nearby streams and the air is polluted with heavy metals and dioxins released by burning waste. Such contamination may be extremely harmful to human health, particularly to children (Brigden et al. 2005; Huo et al. 2007).

Reports by the media and by environmental campaigners, foreign and Chinese alike, have emphasized the highly toxic nature of this business (Baldé et al. 2015; Brigden et al. 2005; Rucevska et al. 2015; Wang et al. 2013). The frequent implication is that Guiyu's inhabitants and migrant workers lack much awareness of the damage they cause to the environment and to themselves; that they mostly lack a moral compass, and that economic benefit rules supreme in their life choices. In this chapter, I debunk this rather superficial portrayal. Pollution in Guiyu is indeed severe: the water is undrinkable, small streams are so thick with pollution they hardly flow, and fields are seriously contaminated. But e-waste work, as all those I encountered in Guiyu were predictably keen to highlight, is not always noxious. Focusing only on the most toxic processes of recycling would severely distort and simplify a rather more nuanced picture.[4] As the above ethnographic vignette illustrates, the common stereotype of shrewd, selfish, evil traders and recyclers, who care about nothing but financial gain and ruthlessly spoil the surrounding environment and locals' health, is partial and misguided.

In order to grasp the complexity and diversity of e-waste work, more attention is required to the historical, social, political, and economic contexts of its development. In what follows, I provide some background on the historical development of e-waste work and its wider regional, national, and global economies. I show that it involves a wide spectrum of activities and occupations, with uneven gains and losses, and a similarly uneven distribution of environmental health harm. This underpins the often conflicting perceptions of e-waste among the local community. Indeed, the conceptual and material ambiguity of e-waste as both waste and commodity lies at the core of how workers and local communities perceive e-waste. The term "scalvaging," a fusion of scavenging and salvaging, conveys the mixed attitude and mixed identity of the materials and goods recycled as both waste and resource (Kirby and Lora-Wainwright 2015b; see also Lepawsky and Mather 2011). To uncover these multifaceted attitudes and practices, I provide some detailed descriptions of the experiences, views, and values of those involved in e-waste work, and the ways in which family and broader social relations intersect with it and facilitate it. This illustrates that even residents of a notoriously noxious place like Guiyu, who—by virtue of having established themselves as one of the e-waste work hubs par excellence—are so frequently blamed for pollution, put forward counter-narratives of relative harm and morality that excuse and legitimize their practices (see Lora-Wainwright 2016). Given the huge diversity of those involved in e-waste work, I provide brief accounts of some very successful businesses, of the poor conditions of migrant workers at the other extreme, and the vast space in between, which is occupied by small, modestly successful family-based processing and trading.

## The Evolution of E-Waste Work: Wealth, Specialization, and Crisis

Guiyu has a population of roughly one hundred and fifty thousand, more than one hundred thousand of them migrant workers from the surrounding area, as well as other parts of China. More than 80 percent of local families are involved in e-waste businesses. Although reliable statistical data on e-waste trade and processing in China is virtually nonexistent given their murky legal status (Tong and Wang 2012; Wang et al. 2013), estimates suggest there are over five thousand family workshops processing between twenty and fifty million tons of e-waste per year (Chi et al. 2011).[5] Due to its lowland morphology, the area is prone to flooding, making agriculture an unreliable source of livelihood. Guiyu's poor farming conditions were in part the reason why locals turned to alternative livelihoods. The presence of convenient trade routes and a historical experience with trade allowed the development of waste recycling since the 1950s, and after economic reforms this became a vital source of livelihood (Zhang 2009, 982; the author's interviews with local residents, 2013).

Juanjuan's father, Uncle Tao, a disarmingly friendly and honest man in his forties, explained that the town had been very poor until recently. In the 1940s and 1950s many locals died of starvation. The situation did not improve much after the founding of Communist China in 1949, and in the 1960s and 1970s many tried to escape by swimming to nearby Hong Kong. Some drowned trying, and being caught would entail incarceration. Uncle Tao's account emphasized the uneven fortunes of the locality and how desperately poor people had been.

With the advent of economic reforms since the 1980s, after the death of Mao, much of south and coastal China, particularly Guangdong province (where Guiyu is located), has become a global manufacturing hub.

Following the pattern typical of this area, large factories sit alongside smaller family businesses run by local entrepreneurs who capitalize on their kinship networks. Indeed, the region surrounding Guiyu comprises several other enterprises, which, those involved in e-waste work are keen to highlight, are also severely polluting. Among them are embroidery, textile manufacturing and dyeing, underwear factories, and toy factories. The toy factories also serve as one of the primary recipients of recycled plastic and microprocessors from Guiyu. Rocketing land prices are evidence of the economic success of the region. In a fully urbanized "village" in downtown Guiyu—where all farmland has been occupied by real estate development—a home of one hundred square meters costs millions of yuan, an unimaginable sum for a village anywhere in China. In nearby towns, in the early 2000s a square meter of land cost a few hundred yuan, but ten years later the cost climbed to eighty or ninety thousand.

The economy of the region, which forms the foundations of China's economic miracle, thrives by exploiting a vast underclass of migrants as wage laborers who have little job security, are regarded as uncivilized and backward outsiders, and are quickly dismissed when business is slow (Lee 2007; Pun 2005). Sprawling towns like Shenzhen (formerly a village, now a massive manufacturing axis) have become magnets for young men and women from poorer rural regions in search for work. Likewise, Guiyu's growing reputation as a center for e-waste dismantling attracted migrant workers who in turn provided the cheap labor force needed to fuel the industry's success.

According to Uncle Tao, the e-waste processing and trading developed gradually. "In the past, we got all this rubbish for free. Then more and more of us became involved in this business. Foreigners realized we made money so they started to charge us. Say I was the first to take waste; I may even be paid to take it. Then my neighbor might find out, and they may go to the US to get it, and then they [traders in the US] would start charging. And the price would grow. To begin with, it's waste! But now we pay for it." At first, much of the e-waste processed in Guiyu came from abroad and seemingly evaded repeated efforts to regulate its flows given its key role in the local economy and government's tax revenue (Tong and Wang 2012). As domestic consumption has grown, however, so has the influx of domestic e-waste that increasingly sustains Guiyu's e-waste-related activities (Minter 2013).[6]

Through the three decades or so in which e-waste processing developed in this area, it has become highly specialized: each village is devoted to different parts of the process and to different components—some streets are covered in signs for microchips, others advertise widely for several types of plastic, yet others offer "circuit board baking" (*shao ban*)—a technique by which circuit boards are heated so that the valuable parts, particularly metals, can be extracted or melted off easily. Having witnessed Guiyu's extraordinary success, nearby towns have also become increasingly involved with e-waste and plastic trade and recycling (see figures 5.6 and 5.7).

Given the hardship they had experienced, even in recent memory, locals enthusiastically embraced e-waste as a livelihood opportunity without much awareness of the pollution it could cause:

When we first started, not a single person in the older generation knew that plastic would pollute; you cannot blame them. People did it to survive. If you had told them from the start how big the impact on the environment was they would not have dared to do it and the problems we have now would not exist. Now everyone is doing this business, and all of a sudden you tell them to stop: it's impossible. ... The

**Figure 5.6**
Discarded air conditioners, 2012.

**Figure 5.7**
Metal parts, disassembled and ready for trade, 2012.

biggest responsibility is on the economy, for the sake of development we sacrificed the environment, and what's more we don't have skills, we can only work at the low-end.

—Li, a man in his twenties, operating a plastic processing workshop, 2013.

As this account suggests, the current heavy reliance on e-waste-related activities as a source of livelihood makes it unlikely, if not impossible, that locals would abandon it overnight. Divergent interests and incentives by central state authorities (which promote environmental protection) and local governments (which prioritize economic growth) produce a fertile middle ground for e-waste work to continue. Local entrepreneurs exploit the sporadic implementation of environmental protection mandates, which is the key to their economic success. Businesses most often operate with the overt or covert approval of local government bureaus, which draw extensive tax revenues from e-waste-related activities. Heavy dependence on e-waste work and the local government's collusion with e-waste businesses go some way to explain locals' defiant attitudes toward attempts to regulate their activities. Taking a diachronic perspective on e-waste work in Guiyu reveals the contrast between poverty in the recent past—still very vivid in people's memories—and the recent meteoric economic rise of the region. This shift powerfully informs current attitudes to e-waste work and the complicit attitude towards its potential harm.

While policies and regulations failed to bring e-waste trade and processing under control, the global economic crisis of 2008 delivered a severe blow to local businesses. All interview subjects, without exception, talked of the negative effects of the crisis. Those trading raw materials were hit most harshly. As capacitor recycler, Guo, explained: "In 2008 the price of copper declined by 60 percent, goods that were worth one million yuan became worth only 300,000 yuan overnight, so traders didn't want to sell. If they did, their savings decreased, so those who couldn't resist went bankrupt" (author interview, June 2013). PC dismantler Linge further elaborated that the entire business was a chain, so if raw material traders were unwilling to trade their goods, other recyclers (like him) were also affected.

Renewed efforts to curb e-waste trade and processing in Guiyu in 2012 and 2013 deepened the effects of the crisis. These were part of broader efforts by the Chinese government to loosen the informal sector's grip on recycling, and to formalize such processes and centralize them in large, state-certified complexes (Tong, Li, Tao, and Cai 2015; see also Tong and Wang 2012; Tong and Yan 2013). As Schulz (2015) very aptly argued, these efforts were not purely driven by a sustainability agenda, but also by political and economic incentives.[7] The Ministry of Environmental Protection (MEP) banned metal refining and severely polluting plastic waste and emphasized the need to close or move polluting industries, upgrade recycling facilities, and stop the flow of illegal waste (Ministry of Environmental Protection 2013b, 32; Ministry of Environmental Protection 2013c). Guangdong province responded by issuing further demands to the municipal level to reform e-waste dismantling and establish a circular economy park in Guiyu (China Environment Net 2012; Shantou City Environmental Protection Bureau Net 2013). TCL—one of China's biggest consumer appliance manufacturers—apparently invested in the industrial park, but, until late 2013, the park still had not been established (see Lora-Wainwright 2016).[8] Uncle Tao was skeptical of this and similar plans, and regarded them as just another plot by the local government to expropriate and sell land: "They just hang a banner, but they don't have the right equipment."

Epidemiological studies carried out in the area may serve as an objective measure of the two periods of slowdown caused by the economic crisis (2009) and by severe crackdowns (2012). Professor Huo Xia has been measuring lead content in Guiyu children's blood for several years. Her data showed a sharp decline in 2009. Indeed, levels were so low that her research team at first thought they had made a mistake. In fact, at that time many workshops closed and thousands of migrants returned to their hometowns, some of them famously riding their motorbikes all the way back to Sichuan. Blood lead levels increased slightly in 2010, and then decreased again in 2011, and even more in 2012 when the more severe restrictions came into force (my personal communication with Huo, June 2013).

Another sign of growing regulation and monitoring was the presence of a privately owned "environmental testing" office. Opened in 2010 in collaboration with a Shenzhen-based company, this office offered a certification service, to ensure that products complied with international standards as required by potential buyers. Local workshop owners brought samples of their products (including plastic pellets, electronic components, and metals) to be tested for quality assurance. Tests cost thirty yuan, and the office provided customers with a full report and a certificate to show to potential buyers. According to the soft-spoken middle-aged man who set up this office, large companies were wary of fakes, but this certificate gave them peace of mind. He also explained that testing was a self-regulating exercise: only local workshops that were fairly confident their products would meet the standards would subject their products to testing. "Otherwise, why would they come to test? It would be a waste of money!" In other words, some local businesses no doubt continue to produce substandard material, but quality assurance is a growing concern. Ironically, this concern with the low quality of Chinese manufacturing influenced his own choice when he purchased the testing equipment. He invested 480,000 yuan for foreign-made equipment—instead of buying Chinese-made equipment for a fifth of the cost—because he feared the tests would not be accurate (and therefore would not be trusted by buyers).

As a whole, the evolution of e-waste work over the past three decades is characterized by the town's growing dependence on it as a foremost means of livelihood. As e-waste work developed, it became more specialized and more integrated within local family, social, and entrepreneurial networks. E-waste work has grown to be inseparable from family life, both spatially—homes and workshops overlap—and socially, as businesses rely on family collaborations and support, and conversely families depend on e-waste for their livelihood. The frequent collusion of e-waste work and local government interests further enabled the former to weather regulatory and economic pressures. Because of its deep embedding within local life, e-waste work in Guiyu has shown remarkable resilience until recently. Such a symbiotic relationship between e-waste work and the local social, political, and economic context powerfully affects local attitudes to the potential harm of e-waste work, and it may explain the sometimes ambivalent, yet resigned, attitudes of locals toward the pollution it may entail. Of course, such attitudes differ across the socioeconomic spectrum. It is to these socioeconomic differences that I turn to next.

## Disaggregating E-Waste Work

Given that the vast majority of e-waste trade and processing in Guiyu is run as family firms, many locals benefitted directly from this lucrative business. Even more than in Baocun and Qiancun, however, benefits were unevenly spread and concentrated in the hands of powerful local families. In Guiyu, some who opened family-run e-waste processing businesses have become unimaginably wealthy. By most accounts, even the wealthiest and most successful families were poor a few decades previously. As in Qiancun, the first to start were most likely to make a fortune. Family connections and entrepreneurial networks—particularly with local government officials, but also with traders, brokers, and suppliers of material—were instrumental to their success. Such good connections and considerable accumulated capital ensured their ability to weather cycles of low trade and to evade crackdowns. For instance, to circumvent regulation, these families may be able to relocate the most polluting steps in e-waste processing to neighboring towns, without actually losing business.

The ability to send children abroad for university education was regarded as a sign of success. One workshop boss, for instance, had sent one son to the UK and another to the US. By far the most conspicuous sign of success, however, was embodied in the construction of five- or six-story homes, preferably on the family's ancestral land, as is common in this region (see Chan, Madsen, and Unger 2009). Those people who could not afford them derided these lavish homes as "face projects" (*mianzi gongcheng*): an exercise in making the family's economic success visible. This in turn further cemented these families' power, reputation, and connections. I was able to visit some of these homes during fieldwork. One of them was the property of the successful boss of a plastic processing workshop, whose daughter was Juanjuan's former schoolmate. When the house was completed in 2007, it was surrounded by farmland, but now several other homes (albeit not as high) encroached upon it. As Juanjuan and I approached, the smell of burning plastic was overpowering. Her friend opened the solid metal door and invited us in. We entered a large (probably two hundred square meters) empty room, with high ceilings, probably conceived as a workshop or storage/trading space and parking space. As we climbed the wide staircase, I glanced across to a lonely treadmill placed at the center of the empty room on the second floor, then saw an equally empty third floor, until we finally reached the fourth floor, where the family lived. Over our head, two more floors lay empty. Wary of not drawing spurious parallels with medieval Europe, I still couldn't stop thinking of San Gimignano's towers in Tuscany, built by lords competing to show off their economic prowess. The living room was extravagantly furnished with a large lacquered and hand-carved wooden couch and four armchairs worth a hundred thousand yuan. A wide, low table sat in the middle of the sofa set, topped by a fancy fine bone china tea set. A large jade sculpture of a sprouting cabbage (a homonym of *facai*, or becoming wealthy) decorated another nearby table. The TV—probably measuring more than one by two meters—was the largest I had seen until then (a few days later I saw an even bigger one, in the house of another plastic recycling family). Juanjuan's friend estimated they spent millions of yuan on this house, but they now considered moving to avoid the ever enchroaching pollution.

Indeed, the ability to afford housing outside Guiyu was an additional sign of success. For many, this was a financial investment, given the growing cost of real estate. But it was also a way to escape pollution. A famous family of four brothers (one of whom was married to Juanjuan's aunt), which was one of the first families to become involved in plastic recycling twenty years previously, serves as a good example. By 2012, they owned a large house in Guiyu where they spent time during the day, were renovating their ancestral home (Juanjuan's father was employed as a worker) also in downtown Guiyu, spent the evenings in a home they purchased in a cleaner nearby town, and owned property in Shantou and Guangzhou.

Their family was unique according to Juanjuan: to maximize on their strong economic interest, the extended family (including grandparents and grand-children) maintained joint ownership of their assets. Each brother (in their thirties and early forties) managed one of the family's four major plastic recycling businesses. Except for one brother and his family who lived in Shenzhen, the rest of the extended family shared all their houses. They employed a maid (for 2,000 yuan a month) who cooked for them all at their main Guiyu home. The business managed by Juanjuan's aunt (shredding plastic and doing injection molding) mostly relied on machines and employed only ten workers, but another of the family businesses—a plastic pearl factory—relied more heavily on manual work, employing more than one hundred workers. She spent most of her time in a small office carved out of a corner of the factory, keeping a watchful eye on the live-stream of CCTV cameras positioned around the factory. An example of interweaving family networks and work opportunities, one of her employees in 2012 was her eighteen-year-old nephew (her older sister's son), whose own immediate family was less well off, and who, by the following year, had also begun to take part in the plastic trade.

While Juanjuan's aunt's family was wealthy, their affluence paled in comparison to families who owned circuit board "baking" and metal refin-ing establishments. As I mentioned above, Lindi estimated that the equip-ment cost hundreds of thousands of yuan, but owners could earn their entire investment back in one or two months. The wealthiest of all, it seemed, were either close to local officials or were local officials themselves. Not all of their wealth was derived from e-waste—selling village land to developers and retaining part of the proceeds was a major source of reve-nue, as is well documented in other areas (Cai 2003; Guo 2001). One Party Secretary, for instance, was said to have bought ten *mu* (6,667 square meters) of land in Shenzhen for more than one billion yuan. Unsurprisingly, eco-nomic and political power frequently overlapped. Lindi and his brother Linge, who recycled PC boards, told me that sometimes those who run metal refining workshops will resort to their powerful connections among local officials, and even hand out bribes in the form of cash or expensive cigarettes, in order to secure their protection and avoid crackdowns. Some of these powerful bosses become the stuff of legend. They were regarded as fearless, physically strong, and socially and politically untouchable. The manager of a metal smelting plant was said to have fallen into an acid tank (used to extract and refine metals) and to have emerged unharmed. One village Party Secretary, a man in his fifties, owned the largest copper plant in Guiyu, a large hotel, and a supermarket. Rumor had it that all his sons and daughters drove racing cars, and that, when his son married, the bride's family paid hundreds of thousands of yuan and bought the groom a Ferrari, costing over a million yuan. The going rate for women to marry into these powerful families was generally hundreds of thousands of yuan. Lindi and Linge summarized their attitude toward these powerful, successful bosses as a mixture of "admiration, envy, and hate." "Common villagers" like them dreamed of becoming one of these inordinately wealthy bosses. They aspired to emulate their power and connections, but at the same time they despised the corruption that earning such status would involve. Realistically, they knew all too well that they would never make it.

At the opposite extreme of the social and economic spectrum to these mighty bosses lies the vast migrant population, who, as in Baocun, undertook the most harmful jobs. Indeed, by providing cheap labor, they were instrumental to these bosses' success. According to one small workshop manager I interviewed in June 2013, 98 percent of workers in dangerous conditions were migrants. For some, wages were not much higher than in less health-threatening occupations. In 2005, they were paid 2,000 yuan per month—a large sum for a young man from rural Sichuan, but not much more than my young friends from rural Langzhong (Sichuan) made in Shenzhen that same year. By 2012, some salaries had risen to 4,000 or 5,000 yuan, but varied widely depending on the job. For instance, older women (the least valued and lowest paid members of the migrant workforce) made 3,000 yuan a month dividing plastic (which is regarded as light work). For migrants, these opportunities were appealing because they had limited options: "The main thing is here they don't require education, only strength; they are happy to employ middle-aged workers [unlike factories]. To work in a big factory you need education, and I'm old, and I don't have an ID, so I don't meet their requirements" (middle-aged man, June 2013).

Locals without the means to establish their own businesses, but with good enough connections to find work, sat one step up on the proverbial ladder. They were more likely to be employed in lighter, less dangerous tasks, such as operating plastic cutting machines and raking plastic pellets. They might earn 3,000 to 4,000 yuan per month for this work, but unlike migrants, they did not have to face the costs of rental accommodation. These salaries were not dissimilar to those offered by other local manufacturing businesses. Men who could not (or were unwilling to) gain employment in e-waste work were employed on the countless huge construction sites fueled by burgeoning real estate development. Much real estate investment itself, as I explained above, was a side effect of e-waste work: successful workshop managers bought land to expand their workshops and build new homes in more salubrious environments. Textile manufacturing, textile dyeing, and underwear manufacturing were alternative, popular options, especially among women. Computerized embroidery was also on the rise. These jobs tended to be regarded as more strenuous than e-waste work, however. They demanded longer working hours and offered less flexibility. For instance, Binbin, one of Juanjuan's former schoolmates, worked in a local workshop selecting microchips. She was paid a meager 1,000 yuan per month, but preferred this work to being a seamstress. (Just to underscore the inequalities inherent in e-waste work: while I talked to Binbin, the workshop owner sold a small bag full of microchips, probably weighing less than two kilos, for 10,000 yuan).

A further step up the ladder was the hugely diverse socioeconomic group comprising traders, owners, and managers of small, family-run processing workshops. During slow periods, those who ran their own e-waste businesses might not make much more than the workers (only a few thousand yuan). However, when business was thriving, they might earn as much as one hundred thousand yuan a month. Income, of course, also varied vastly depending on the nature of trade and processing, with metal refining and circuit board baking occupying the top of the ladder. Since describing adequately the huge variety of processing workshops is beyond the scope of this chapter (or indeed an entire book), I will limit my account to two cases. First I will turn to Juanjuan's family.

In 2012, Juanjuan's family was composed of nine people: Juanjuan; her twenty-year-old brother; her twenty-three-year-old sister; her twenty-five-year-old brother, his wife, and their newborn twins; Juanjuan's parents; and her paternal grandmother. Their home was located in a small village just outside Guiyu town, and was still surrounded by paddy fields, although Juanjuan explained that there used to be much more farmland before the village expanded. Her house was a beautifully decorated traditional courtyard home typical of the Chaoshan area (see figure 5.8). Opposite the main

**Figure 5.8**
Juanjuan's home, 2013.

entrance were the kitchen, toilet, and shower room, and just to the right of the entrance was a small dining room. A second entrance led into the open courtyard, which in turn led directly to the main living space, including the ancestral altar. This room was furnished with a polished wooden sofa and chairs, a low table, and a large flat-screen TV. Hanging on the walls were some watercolors of peonies and a large digital clock with a moving image of a waterfall surrounded by verdant hills. On the main wall facing the entrance to the room was a long, wooden cabinet that occupied the entire low part of the wall. On top of it was a wall-to-wall decorated mirror and above it a row of small frescoes. Bedrooms, as well as the traditional family hearth (not in use, but beautifully kept) were arranged along two sides of this room. The house had been redecorated twice: first when Juanjuan's parents were married, and again two years previously, for her eldest brother's wedding. In her village, these "traditional" homes were still in the majority, but they were fast being replaced by highrise blocks built to exude financial achievements.

Several members of Juanjuan's extended family were involved in e-waste trade and processing—her aunt had married into a successful plastic processing family, her cousin was a plastic trader, and his fourteen-year-old sister stripped earphones on weekends for forty-seven yuan a day. Her immediate family's engagement with e-waste work was uneven, however. Her father, Uncle Tao, had been until recently the main earner, and his employment had been patchy at best. In part, this was due to severe illness in the family: his youngest son was diagnosed with a brain tumor as a child and had undergone several rounds of very expensive treatment that had swallowed up much of the family's resources. This pushed Uncle Tao to experiment with several types of work, in order to cover his son's healthcare costs. Attached to one side of their courtyard house was an annex built with bare air bricks, probably measuring around one hundred square meters. When it was first constructed in the early 2000s, it hosted a small underwear factory set up by Uncle Tao, but this business did not turn out to be lucrative. Inspired by developments in nearby Guiyu, Uncle Tao was one of the first in his village to start a plastic recycling business, but this too was short-lived. For several years, he worked in a local quarry co-run by a relative, but he lost his job in 2009, just when his younger son was undergoing the latest course of treatment in Beijing. He recalled finishing all their savings and borrowing considerable amounts, especially from his sister who had married into a successful plastic processing family. In the intervening years, Uncle Tao continued to find short-term employment: in 2012 he worked as a builder for his sister's family; and in 2013 he found employment in another quarry, which required him to live away from home during the week.

Inspired by his father's efforts with e-waste work and by its growing prominence, Juanjuan's eldest brother had made several attempts at establishing successful businesses. As seems common, he had swung between processing and trading, employing the connections, skills, and knowledge gained in one business to help with the other. He started out by trading plastic, and for this reason he had rented the workshop in downtown Guiyu where I stayed with Juanjuan in 2013. He had lived there with his wife, but they moved back to the village in early 2012 after she became pregnant. As trading went through a slow period, Juanjuan's brother turned to plastic processing—one of the sectors of e-waste work least affected by the economic downturn and the cycles of regulatory crackdowns. Together with his cousin (who had worked in his aunt's plastic processing factory) they purchased a plastic press, which was stored in the small warehouse attached to Juanjuan's home. Juanjuan's younger brother also worked for his older brother when he started to do plastic cutting. Having been a trader, her older brother had the necessary contacts to secure material for his plastic press. He also had contacts that would assist in selling his products. This business too, however, had come to a temporary standstill. They had not yet sold any of their products, and Juanjuan's older brother had set up yet another business doing computerized embroidery, this time with his wife's relatives.

Juanjuan's family's involvement with e-waste work displays some of the patterns of collaboration and support that are common in Guiyu. Relatives may pool resources to establish new businesses; a son may pick up business where his father left off; relatives may be employed when business is going well and left idle when it is not. More successful relatives may help out. Their case also illustrates that those like Juanjuan's older brother may shift between different types of e-waste work, drawing on their existing networks and knowledge, and experiment with different lines of work, especially when the current business is failing to bring in significant returns. Finally, it shows that illness can be a major drain on resources (see Lora-Wainwright, 2013a). In these cases, success in e-waste work among one's extended family may come to the rescue. This surely gives a very different connotation to e-waste work—not so much as a threat to health but as the means to protect it.

Capacitor recycling occupies a prominent place as a fairly financially secure and relatively clean business. It was particularly common in a town bordering Guiyu, where there were more than a hundred capacitor recycling workshops. Guo and his wife Peng—a couple in their mid-twenties with two young children—ran a successful family workshop for about six years in downtown Guiyu. They lived on the third floor of a five-story home and rented two floors in the building opposite, where the workshop was situated. Guo stopped attending school at seventeen and worked in Shenzhen for two years, gradually learning some basics of e-waste work. Peng worked in textile manufacturing before marrying, and now helped her husband to oversee workers. Guo first set up the capacitor recycling workshop with a friend, but by 2012 he ran it with his wife, taking the responsibility for trading goods. Six CCTV cameras were installed in the workshop and connected to the TV screen in their home, from which Peng could keep an eye on workers remotely. When business was good, they employed twenty workers, but in 2012 they only had ten. All workers were young, most under twenty, and some were migrants from Guangxi.

Guo regarded his work as not very lucrative, but also relatively light (*qingsong*). He explained the rationale for recycling in financial terms: a new capacitor costs a few yuan, but an old one is only one yuan, and the quality is still very good. To underscore this point, he stressed that all the goods he recycled were imported (particularly from Japan, Korea, and Taiwan), and therefore better quality than domestic-made capacitors. The price depended on their serial number (linked among other things to their age) and size (not their country of origin), and ranged from one yuan to around ten. He emphasized that he did not make much profit: if he bought a capacitor for one yuan, he might sell it for 1.1. The real profit, he argued, was made by the buyers. His main customers were TV and PC factories in Shenzhen. When he traded within Guangdong (Shenzhen, Foshan, Dongguan) he went there personally, but he also traded with other parts of China, such as Shanghai, Zhengzhou, Xiamen, Tianjin, and Shandong province.

Most capacitors he recycled were originally part of TVs and PCs. TVs and PCs reached the Guiyu area after passing through Nanhai (Foshan city, Guangdong). Limited processing activities went on in Nanhai itself, due to strict controls. These items were dismantled in workshops in Guiyu or in a nearby town, and Guo bought capacitors directly from them. At a first glance, Guo's workshop seemed messy and rudimentary, with boxes piled everywhere, but it ran as an efficient and well-oiled production chain (see figure 5.9). Capacitors made their way through several hands in Guo's workshop—with each of the workers specializing in a particular part of the process. Upon arriving, capacitors were stripped of their plastic covers and brushed clean. A second employee tested them to establish if they still worked. Faulty ones were discarded and sold by weight to other workshops, which smelted them to extract aluminum. Working capacitors were heated to remove residual parts originally attached to motherboards. They were then positioned vertically on large charging shelves and recharged (see figure 5.10). The next step was wrapping them in their new plastic cover. When we visited in 2012, the cover of choice was a counterfeited Japanese brand. (As I watched over the process I wondered a number of things: If some of these capacitors were originally made by that Japanese company,

**Figure 5.9**
Bird's-eye view of Guo's workshop, 2012.

**Figure 5.10**
Recycled capacitors being recharged, 2012.

did this process still count as counterfeiting? Where are these goods produced? Here? At their original production site? In both locations?) In order to seal the plastic cover, the capacitors were placed on small trays that were put in what looked like a microwave for a few seconds, and then fanned with a blow dryer (see figure 5.11). Finally, capacitors were tested and placed neatly in cardboard boxes ready to be sold all over China.

Guo and Peng were adamant that their work caused practically no harm to the environment or to health. In large part, they were right—with the exception of the two young men who melted metal residues at the tip of the capacitors, in a poorly ventilated room and with no masks, and the two

**Figure 5.11**
Recycled capacitors wrapped in new covers, 2012.

young women who baked and dried the new plastic covers in a windowless room. But when one thinks of life in some parts of Guiyu, with black and smelly streams, acids and heavy metals leeched into the soil, and the over-powering smell of plastic, the effects of Guo and Peng's workshop did rank rather low. In conclusion, I turn to examine how locals evaluated the effects of pollution, how it has become normalized, and how resignation intersects with efforts to deny blame.

## Between Concern and Denial: Normalization, Hierarchies of Harm, and Redefining 'Health'

The presence of pollution in Guiyu was a relatively uncontested and uncon-testable fact. Most of those interviewed—ranging from migrant workers to wealthy local bosses and locals working in unrelated businesses—highlighted its existence. Much of their evidence was put in sensory terms: "the air smells very badly," "the water stinks," "the air is yellow," "the water is black," and so forth. A few interlocutors downplayed the severity of pol-lution, and argued that media reports exaggerated its impact by treating specific, extreme cases as if they represented the entire town. Some were also keen to emphasize that since stricter controls were put in place in 2012 air pollution had decreased. Yet, overall, pollution had become accepted as a fact of life; it was only a matter of degree.

Responses became much more varied when I raised questions about pollution's effects on health. Several health conditions were regarded to be prevalent locally and considered to be linked to pollution. Nose infections, coughs, respiratory tract problems, and darkened teeth were mentioned most frequently, followed by vulnerability to flu, headaches, poor hearing, skin problems, lung problems, and liver damage (including cancer). How-ever, responses ranged widely from confidence in the harm of pollution to denial of any demonstrable link between pollution and disease. Such differ-ences may in part be correlated with socioeconomic differences, but not in any simplistic way. Conversely, in cases where a migrant and the manager of a successful workshop voiced the same response, its implications and connotations would be predictably different.

Migrants rarely expressed confidence about the harm of pollution, but rather they were either unsure or openly skeptical about it. They cited the presence of locals in their eighties and nineties as evidence that pollution did not pose a significant threat. One person disputed the prevalence of ill-ness: "I heard about children being diagnosed with illnesses, but I don't know if it's true. ... I also heard that the soil and rice are polluted, but every-one is in good health." Another migrant accepted that there were sick peo-ple, but he questioned whether this could be attributed to pollution by arguing that there was no proof of a link and that people suffered from the same illnesses even in areas where the air was clean. Such skepticism was not only confined to migrants. Indeed, a local e-waste worker and a migrant strikingly used the same expression in two separate interviews: "If everyone was sick, who would still dare to stay here?" Like migrants, locals who denied the health effects of pollution resorted to a range of examples as evidence. They referred to the presence of many villagers who lived to a ripe old age and argued that illness may be linked to genes and lifestyle. One stated: "Everyone's body is different. You can't say it's all due to the air [pol-lution]. Some people are just unhealthy anyway." Skepticism and the emphasis on the uncertainty surrounding causes of illness served as a sort of self-defense mechanism, enabling the residents to reassure themselves that living and working in Guiyu was not causing any lasting damage.

A crucial strategy to downplay pollution's effects on health involved putting their correlation itself under scrutiny. A local manager reflected: "The air is bad, many have nose infections, unhealthy lungs. But it's hard to say [if it is connected to pollution]. Some have nose infections, some don't. ... Those who burn circuit boards may harm their liver, but it is hard to gather evidence of whether the liver function is poor. These illnesses exist everywhere." This short statement questions the effects of pollution in three ways. First, the intimation that not all in Guiyu suffer from nose infections implies that air pollution may not be the cause. As the reasoning goes, if air pollution caused nose infections, then everyone should be affected. If some people were not, then something else must be to blame. Second, and conversely, he suggested that these illnesses exist in areas not affected by pollution, and therefore pollution cannot be the sole culprit. Third, he added that, at any rate, evidence is hard to find. His points are, of course, scientifically accurate. Epidemiologists would not conclude that the co-presence of pollution and illness proves that individual illness episodes must be due to pollution. Indeed, epidemiological studies never focus on single cause-effect relationships, but on correlations that involve a wider range of factors. There can be little indisputable proof, on an individual level, that a single factor (pollution) caused a particular health effect (nose infections or liver damage). But this manager's insistence on the uncertain link between pollution and illness is significant not so much for its scientific accuracy as for its social, political, and economic contexts and implications. It is a sign of how intimately embedded pollution is in the lives of Guiyu's people and a powerful strategy that enables them to question the effects of that pollution.

As testament to the normalization of pollution in Guiyu, locals often swerved around the question of impact. Two research participants responded in remarkably similar ways to questions about the effects of pollution on health: "Why would you worry about that? Do you have nothing else to do?" Another replied: "Who is so wealthy that they can worry about that?" It is important to note that these statements do not deny that pollution may indeed cause illness, but rather they voice a refusal to actively consider these effects, resembling Guo Lin's comment in Baocun that "it is best not to think about things you cannot change" (see p. 86). For those who could not escape pollution, denying its effects served to minimize the emotional and psychological strain caused by accepting that the environment may be harming not only them but also their families.

Virtually every family has at least one member who is involved in e-waste work. The resulting normalization of pollution may be encapsulated by the frequent remark: "Everyone here does this [e-waste work]." One local poignantly added: "Opposing pollution is just like opposing yourself" (*fandui wuran jiushi fandui ziji*). Another explained that: "Nobody complains. Everyone wants to have a good life for themselves. They don't want to provoke others." Locals explained that, typically, those who operated workshops would take care not to impact too much on neighbors. Indeed, while the increase of emissions at night is often seen as an effort to avoid crackdowns, some locals also regarded it as a way to decrease the effect on neighbors: at night, people are more likely to be indoors, with their windows closed, and to thus be less affected by air pollution. As one local put it: "If you can endure, you endure. As long as it's not too excessive [*guofen*] it's fine. Everyone hopes they can have friendly relationships. We take these relationships very seriously." The high value placed on social relations therefore superseded concerns about the potential harm caused by pollution. Tolerance toward pollution caused by other locals is nothing surprising and has been documented elsewhere in China (see chapter 2). In extreme cases, Uncle

Tao told me, villagers may call 110 (China's equivalent of the US 911). But the fact they would only do this in "extreme cases"—when the air is so smelly it becomes hard to breathe—is telling of quite how low their expectations about a healthy environment have become.

Resignation to pollution as part of the status quo did not always result in skepticism about its effects. Some locals and migrants were quite confident about its harm. They tended to be younger and better educated, but not exclusively. Asked whether pollution had any impact locally, one young migrant from Guizhou, who had worked in a plastic processing workshop in Guiyu for six years, argued: "Of course it has a very big impact. It is *because* pollution here is too severe that many have nose infections and lung diseases. ... Processing circuit boards is quite poisonous, the skin is quite sensitive to it" (my emphasis). In his account, a causal relationship between pollution and health effects was stated openly. A local university student specializing in medicine elaborated further: "If you researched it, you would find many have respiratory system problems. Taking apart circuit boards is very harmful to health. Locals rarely do it; they mostly employ migrants to do it." This student presented the fact that locals refused to take on these jobs as evidence of their harm. A university graduate (in international marketing) in his forties, who had recently returned from Beijing, was confident that pollution was very harmful and cited research reports he had read to support his argument. He also cited the inability of men to pass the army entrance test in recent years as evidence that locals' health was severely affected. Finally, capacitor recycler Peng cited her personal experience with her son as proof of the adverse effects of pollution: "Look at my child! He is often unwell. When I take him to hospital there is a long queue of children with a sore throat and a fever."

Tellingly, all those who stressed a link between pollution and illness—capacitor recyclers Guo and Peng, CD drives recycler Lindi, the young migrant from Guizhou, and university graduates and students—were also keen to emphasize that their line of work was not responsible for pollution. For the most part, this was empirically true. But it was also a powerful discursive strategy through which these locals were acknowledging the severe and yet normalized presence of pollution while at the same time emancipating themselves from direct blame. Accordingly, they critiqued local corruption that allowed some to benefit unimaginably from polluting work, and they advocated for stricter regulation of the most polluting activities, but defended their own work as relatively harmless and even "environmentally friendly."

These efforts by locals to discriminate between different types of e-waste work and to carefully position themselves on a moral footing show that e-waste work may not be treated as a monolithic whole. Its huge diversity enables some to refer to hierarchies of relative harm to highlight their limited culpability. This significantly complicates any simple narratives of victims versus perpetrators. At the same time, the long evolution and local embedding of e-waste work in Guiyu has produced a pervasive sense of resignation to pollution's presence. This fuels uncertainty over the environmental health harm of e-waste work and supports processes of embodied attunement to pollution. Unlike in Qiancun and Baocun, the acute awareness of the interdependence of each aspect of e-waste work and trade on the entire system keeps local residents from complaining. The vast disparities in economic gains across e-waste workers further undermines the formation of a collective sense of shared environmental health harm. Those who can afford to do so protect themselves from harm by moving elsewhere, by sending their children to study in other towns, or by at least buying bottled water or filtration systems. Those who cannot move endeavor to at least protect themselves discursively: they lower their expectations for a healthy environment and they uphold the internal diversities inherent to e-waste work in order to avoid responsibility for pollution and to emphasize their relative distance from its worst manifestations. Resigned activism in Guiyu hangs in this delicate, uneven, and shifting balance between praising the recent economic boom and despising the unequal distribution of gains and harm; between resignation to pollution and resentment toward it; and between the denial of pollution's harm and the denial of blame.

# Conclusion

## Comparing the Sites

At the end of chapter 2, based on an overview of cancer villages studied by Chen and his team, I extrapolated a set of factors which influence individuals' and local communities' attitudes and responses to pollution and their effects. Let me now return to those factors to draw some comparisons across my three case studies and to consider how they shape the focus of complaints, activist strategies, and the effects of activism. In turn, this will inform some reflections on the various limitations to the role Chinese villagers can play in environmental protection. Table 6.1 should provide a schematic bird's-eye view.

### 1.  Types and Levels of Pollution, Relative Clarity of Its Link with Particular Illnesses, Level of Awareness of Pollution, and Its Harm

Industry, mining, and recycling activities in the three sites have seriously degraded local ecosystems; they affected the soil and water sources, and made them unfit for farming and other forms of provisioning in the most polluted areas. While the types of pollution differ, as do their effects, the severity of pollution is comparable and in all three cases has affected the locality for decades (although in Qiancun it has declined since the mining ban). Across all three sites, just as in the cancer villages studied by Chen and his team, residents became aware of pollution in similar ways, first and most importantly through direct experience, which was later supported by tests on the local water sources (one in Baocun, several in Qiancun and Guiyu), on the soil and crops (Qiancun), and on the air (Guiyu).

In all three cases, pollution is quite clearly correlated with locally prevalent ailments. However, just as in the cancer villages studied by Chen and

**Table 6.1**
Factors influencing rural activism and their effects.

|  | Tacun (Baocun) | Fengcun (Qiancun) | Guiyu |
|---|---|---|---|
| 1a Type of pollution | Phosphorous, sulphuric acid, and other chemicals used in fertilizer production. | Lead and zinc. | Heavy metals, especially lead; acids used to process heavy metals, including aqua regia. |
| 1b Levels of pollution | Severe. | Severe. | Severe. |
| 1c Relative clarity of link between pollution and illnesses | Epidemiologically clear link between phosphorous and fluorosis; fairly clear link between local types of air pollution and respiratory infections. | Clear link between lead exposure, epilepsy, and impaired mental capacity among children. | Clear link between lead exposure and impaired mental capacity among children; clear link between tooth deterioration and locally prevalent water pollution. |
| 1d Levels of awareness of pollution and its harm | High but mixed with uncertainty. | High. | High but some key information is missing; strong sense that harm is uneven for different types of e-waste work. |
| 2a Texture of community | Formerly united, later stratified and divided by industry. | United, single surname, but stratified by mining. | Kinship solidarity, but divided by uneven gains. |
| 2b Organizational potential | Sporadic. | Strong, but not sustainable. | Strong, but targeted at protecting the sector. |
| 2c Presence of charismatic grassroots leaders and their ability to attract attention | Co-opted into the local government or bought off by compensation deals. | Skilled petitioner, but discredited because of his class label. | Powerful, feared figures are typically beneficiaries of the most polluting and most lucrative sectors of e-waste work. |

**Table 6.1** (continued)

|  | Tacun (Baocun) | Fengcun (Qiancun) | Guiyu |
|---|---|---|---|
| 3a Relationship to polluting firm(s), extent of dependence on polluting firms | Symbiotic: some locals are retired workers and registered residents receive compensation. Large SOE: some opposition, but targeted at compensation. Smaller plants: stronger opposition. | SOE mine: strong hostility. Outside contractors: strong hostility, especially since 2008. Smaller mines co-owned by villagers: little hostility. | Symbiotic: most locals are connected to e-waste work; domestic space and workspace are inseparable. |
| 3b Support from different levels of the state | Village cadres: mediators with industry. Township: oppressive. Central state: seen to support industry. | Village cadres: do not take part in petitions and visits. Township: limited role. County: visits, but takes little action. | Village and township cadres: directly benefit from e-waste work. Central state: attempting to regulate and formalize e-waste work. |
| 4a Support from civil society | None. | None. | Not sought, though several reports have been compiled by major organizations. |
| 4b Media exposure | Media interest was rejected. | None. | Media interest is regarded as a threat. |
| 5a Focus of complaints | Initially, health and polluted environment; recently focus on compensation, not health. | Sustained complaints about illness, polluted environment (particularly water), and inadequate compensation. | Environmental effects of circuit board baking and acid baths. |

**Table 6.1** (continued)

|  | Tacun (Baocun) | Fengcun (Qiancun) | Guiyu |
|---|---|---|---|
| 5b Activism strategies | Small blockades. Resort to village cadres as negotiators. Demand compensation instead of regulation. | Petitions. Visits to the county government. Violent protests in 1980s and 1990s. | Local protectionism. Counter-discourses of relative harm. |
| 5c Effect of activism | Compensation trap. | Visits by county government, water tests, but limited redress. | Limited regulation and limited formalization until 2014. |

his team, the relative clarity of these correlations did not play a determining role in whether locals took action, what kind of action, or its effects. They retained a sense of uncertainty about whether they could indeed prove that pollution caused particular health effects, and, in Guiyu, a few research participants lacked some basic knowledge on the health effects of lead contamination. As I have argued throughout, local residents were more likely to discount widespread and relatively minor ailments as inevitable. By contrast, where tests showed the presence of harmful substances in excess of acceptable standards, these tests themselves were taken as evidence of harm. Conversely, it was on the basis of these tests that local residents demanded intervention, such as provision of safe water, and it was on this basis that they were more likely to receive it (for instance, water pipes in Qiancun), however uncoordinated and ineffective these interventions may be in the long term.

## 2.   Community Cohesion, Its Organizational Potential, and the Role of Charismatic Leaders

The extent of community cohesion varied across the three sites. As a whole, development has created increasingly divided communities, where benefits and costs are unevenly spread, and interests and entitlements are similarly uneven. In each chapter, I disaggregated the otherwise monolithic term "local community" to highlight divisions between administrative villages and natural subvillages, locally registered residents and migrants (Baocun), those with more or less strong connections and social networks, early and late starters (Qiancun and Guiyu), and between different types of work (Guiyu). However, the effects of these divisions upon the forms of activism embraced varied depending on the wider political economic context, the presence or absence of kinship solidarity, and the presence of charismatic leaders.

In Baocun, weak kinship solidarity and financial incentives in the forms of land rental fees and compensation packages for locally registered residents underpinned the sporadic and economically oriented nature of collective responses to pollution. If, as many noted, in the past villagers may have united to protect the collective interest, industrialization largely disintegrated their sense of a shared interest and effectively split local residents by virtue of their diverse entitlements. As a consequence, in the present "each protects him/herself" (frequent statement, Baocun, 2009). Small-scale, localized, and routinized protests mounted most frequently by a handful of individuals only resulted in compensation, and this eventually became their ultimate aim. Where potential protest leaders emerged, they were soon co-opted into becoming part of the village council and becoming negotiators between villagers and local firms.

In Qiancun, stronger kinship solidarity supported more resilient collective efforts to demand redress, both in terms of obtaining the right to mine and in terms of curbing pollution. A clearer sense among villagers that the main formerly state-owned mine, and mines contracted to outsiders, reaped the lion's share of benefits also strengthened these efforts. The presence of an educated and resilient petitioner (and a doctor who supported him) further channeled these complaints toward demanding a cleaner environment, although his efforts were undermined by his tainted political identity. Although they seem incapable of sustaining collaborations to tackle challenges like provision of clean water, and their organizational potential proved weak in the long term, Fengcun villagers have continued to complain to the county government about pollution, despite the limited effects of their efforts and the failure of several attempts to provide safe drinking water. These efforts are hampered by uneven resources within Qiancun village, water sources being numerous but unreliable, lack of coordination between villagers to maintain water projects in the longer term, and limited funding at the county level devoted to water provision.

In Guiyu, strong kinship solidarity within villages had distinctly different effects than in Qiancun, due to the different political economic dynamics of the two localities. Where in Qiancun locals competed with the state-owned mine and later with private mines, in Guiyu the entire sector was established and still run largely by locals. This resulted in strong local protectionism rather than in demands to curb pollution. As in Baocun, village and township officials benefitted directly from deals in the e-waste sector, from land rental fees, and, in at least some cases, from managing recycling and processing workshops. These charismatic and often fear-inducing characters further entrenched e-waste work in the locality rather than complain about its effects.

### 3. Local Political Economy, Degree of Dependence on Industry, and Relationships to Various Levels of Government

All three areas were prevalently agricultural before the opening of the mines, processing plants, and start of recycling activities. In Baocun and Qiancun, a resource curse meant that abundance of minerals attracted development. In Guiyu, its convenient position on trade routes supported the development of the e-waste trade. In all three cases, the local economy is overwhelmingly dependent on polluting activities, and the relationship between the locality and polluting activities is symbiotic. All three areas attract a large number of migrant workers (though since the closing of most of Qiancun's mines in 2008, many have left). In all three cases, the development of industry, mining, and processing resulted in increasingly unequal societies, in which some were better positioned to benefit from development than others, creating tensions between social groups, and, in Qiancun and Baocun, between local residents and the main local employer.

The pathways of cost and benefit to the local community differ across the three sites. In Baocun, at first locals were employed as formal workers (high gains). As these workers retired, the following generation of locals were only able to gain employment as menial workers (low gains), while they suffered pollution's effects on their crops and health. Most recently, though pollution has not decreased considerably, compensation packages have increased and land fees were shared among villagers (high gains). In Qiancun, by contrast, villagers were never employed as formal workers and always perceived the state mine as an enemy of sorts, which extracted resources without giving much in return (low gains). Through recurrent fights and petitions, villagers obtained some compensation and the right to mine and sell minerals (high gains). In the long run, however, they continue to live in a contaminated environment and, after most mines were closed, they gained no compensation, could not rely on mining, and were unable to rely substantially on farming the polluted and dried land (low gains). In Guiyu, patterns of cost and benefit to the local community have been more consistent through time. Since the start, locals have been the main driving force and main economic beneficiaries of e-waste processing, though benefits also depended on how early families got involved in the trade, on the quality of their networks, and on the activities they engaged in (plastic cutting, for instance, is less lucrative than metal smelting and refining).

The relationship among villagers, polluting firms, and various levels of the government also varies across the three sites. In Baocun, the main industry (Linchang), as a former SOE, was regarded as a national priority and seen to be endorsed by the central government. The township government's rejection of villagers' petition in the 1980s was also seen as a sign of their support for the industry. The considerable tax revenue that the township derives from local industries was (rightly) considered a further motive for their support for Linchang. The position of the village government has shifted over time and it recently came to play the role of mediator between villagers and industry, securing financial benefits for registered residents and maintaining social stability.

In Qiancun, the role of the central state was mixed: it supported mining, but issued a ban on small-scale mines that impacted adversely on villagers. The county government received regular appeals by villagers, but was seen to have taken limited action in support of villagers, as was the township. Finally, village cadres, unlike those in Baocun, rarely supported locals' complaints, and their role as negotiators was also limited. In some ways similar to Qiancun's small-scale mines, Guiyu's informal e-waste work has recently been the target of severe central government mandated crackdowns and regulatory efforts in an attempt to formalize the sector. By contrast, village governments sided more closely with local residents and sought to protect and sustain local businesses. The township government occupied an ambiguous position—on the one hand supporting formalization, but on the other also benefitting directly from the local informal sector. Such uneven levels of support and opposition across different levels of the Chinese government affected the avenues of activism embraced and their effects.

## 4.   Support from Civil Society, the Media, and Outside Expertise

The effects of civil society support, media exposure, and scientific evidence are complex and depend closely on the other factors outlined above. Neither Baocun nor Qiancun villagers resorted to civil society organizations or to the media to attract attention. Indeed, Baocun villagers turned away reporters on one occasion, preferring instead to protect the locality's economic interest from potential fines for infringing environmental regulations. In both sites, unlike in the cases of cancer villages, which demanded and attracted such attention, contention remained localized and relatively inefficient at curbing pollution. In Qiancun, tests and research carried out over several years have played a dual role: they simultaneously strengthened locals' sense that pollution is severe (even when test results were not disclosed), and they strengthened locals' sense of resignation by virtue of the limited effective intervention despite evidence of pollution. This highlights that scientific evidence does not necessarily result in intervention, and, as I argue in chapter 4, many complex obstacles also affect the shape that intervention may take.

Qiancun's and Baocun's absence from the media sphere stands in stark contrast to frequent and damning reports on Guiyu and civil society interest in the area—both of them largely unwanted by local residents and the government alike and seen as a threat to local livelihood. Whereas my analysis of cancer villages in chapter 2 would suggest that media and civil society attention are crucial to improving regulation, Guiyu's case highlights that these are insufficient if the local population and the local government are determined to protect local businesses. Overall, we might conclude that when media and civil society support are absent, it is unlikely that activism will transcend the locality and be visible or have any sustained regulatory effects. However, the presence of media attention and civil society support from outside organizations themselves do not necessarily translate into better environmental protection unless locals themselves also demand it. In other words, the media and civil society are important forces, but they are doomed not to have a sustained or effective impact unless they are coupled with both resilient complaints by the local population and strong intervention on the part of the state.

## 5.   Activists' Focus, Strategies, and Outcomes

All of the factors elucidated above are crucial in shaping different types of activism, their focus and their effects, and particularly the role that communities can play in enforcing environmental protection. In Baocun, initial complaints and a petition had a regulatory aim, but were met with oppression. As these demands for regulation failed, repertoires of action changed and became more reliant on village cadres as mediators and on small blockades to exert pressure on the village cadres and on the firm. In turn, the initial regulatory aims were largely abandoned and residents became relatively acquiescent toward pollution and content to draw compensation and high land rental fees instead. Dependence on industry, a stratified community, and the relationship among villagers, industry, and local government shaped co-opted forms of low-scale activism. Residents ultimately play an adverse role in enforcing environmental protection because they have become complicit with the polluting firms. Their actions are aimed not at regulation but at compensation. In this context, it is unlikely that villagers in Baocun would be the force demanding an end to pollution, especially if it meant less income. Only strong top-down intervention from beyond the locality could work to regulate industries more strictly.

Where Baocun residents have become relatively acquiescent toward pollution and content to draw compensation and high land fees, in Qiancun they continued to wage demands for pollution to cease. All the ingredients seem to be in place for activism aimed at stopping pollution: villagers have had several tests done over the years, the community was relatively cohesive against non-locally owned mines and processing plants, and following recent restrictions on mining they were left with mining's negative side-effects, but without its benefits. Protests, petitions, and visits to the county government managed to attract government attention, obtain water tests, and win some compensation and the right to mine and sell minerals. However, while regulation was one of their aims, their actions had little effect on that front. The few interventions that have taken place to secure safe water or a cleaner environment have been largely unsuccessful. Local processing plants (widely believed by villagers to be the worst polluters) only closed when mining was regulated more strictly by district directive—not as a result of citizen pressure. Even then, regulation only affected small mines run by villagers; it did not put an end to pollution, nor did it clean up existing pollution. While villagers may continue to put pressure on the local government to address pollution, outcomes are constrained by limited capacity. More sustained and coordinated efforts by the state would be needed to achieve long-term regulation and cleanup. Until then, villagers are likely to continue to feel disempowered and condemned to live with pollution.

In many ways, Guiyu is a classic case of local protectionism: despite intense media attention, civil society interest, and scientific studies on the environmental health harm of heavy metals, local residents were protective of the sector and emphasized their dependence upon it. Village cadres were similarly complicit in e-waste work, and therefore inclined to circumvent regulation rather than enforce it strictly. As a consequence, in Guiyu, collective opposition to pollution was largely absent, and local residents responded to severe pollution by attempting to protect themselves and their families—for those who could afford to do so, by living elsewhere. Given the embedded nature of e-waste work in local society, repeated attempts to regulate and formalize the sector failed to change things substantially.[1] Ironically, perhaps, the most effective regulatory force was the global economic crisis of 2008 and the slow and uneven recovery since then. The plummeting price of metals brought e-waste trade to a temporary standstill. Ultimately, then, in Guiyu, as in Baocun, citizens play an adverse role in enforcing environmental protection because they have become complicit with the polluting firms. Only strong top-down intervention from beyond the locality could work to regulate industries more strictly.

As a whole, the three case studies suggest that villagers' role in decreasing pollution is limited in all three sites, but for different reasons. In Baocun, villagers have been co-opted by compensation packages and shared land rental fees that prompted them to accept the presence of pollution. In Qiancun, locals have maintained more resilient demands to curb pollution, but limited capacity at the county level resulted in limited interventions. In Guiyu, the close involvement of most members of the local community in e-waste work created incentives to protect it.

**Lessons for the Study of Chinese Environmentalism**

Several conclusions may be drawn from these examples. First, in rural settings highly dependent on polluting firms, and in the absence of media or NGO support, collective action is mostly limited to direct negotiations with the firm, petitions, appeals to local state regulators, and small, sometimes violent, protests. Second, villagers are less likely to oppose pollution when it may be economically damaging to them to do so. Third, a cohesive community and the presence of charismatic leaders are instrumental in supporting the development and success of collective action. Conversely, socioeconomic stratification and diverse entitlements fragment the local community and undermine the development of collective action. Fourth, activism to demand redress, even in its most contentious and confrontational forms, can only succeed if the state intervenes to play a regulatory role, and if the local government has the capacity to implement coordinated and sustained interventions.

In recent years, China's environmental activism has seemingly grown in intensity and become more visible in media reports.[2] Based on a study of four urban and peri-urban cases of environmental contention between 2007 and 2013, Steinhardt and Wu (2015) have proposed that a new repertoire of action is developing that involves closer linkages between protests and NGOs/policy advocacy, broader protest constituencies, mobilization "in the name of the public," and a focus on preventing projects rather than ex-post-facto protests. They acknowledge that, since the new national leadership took office in 2012, constraints on civil society have intensified, causing the innovations they identified to become less prominent, but they maintain that such innovations "are likely to rebound once an opportunity presents itself" (82).

The changes highlighted above were not prevalent in the three cases examined in this book. Why might that be the case? First, repertoires of contention and forms of capital (economic, social, cultural, and political) available to rural communities in opposing environmental harm are much more limited and less likely to intersect with urban-based ENGOs. Second, while Steinhardt and Wu describe contention against planned projects, in the cases examined in this book pollution has already affected the locality for decades. This produces very different dynamics. Polluting industries have become embedded in each local context. In a sense, then, the pattern of resigned activism is more prevalent when it comes to responses to existing pollution. Third, although the importance of public participation is highlighted as part of the government's ambition to build an "ecological civilisation" (Geall 2015), social stability remains a vital focus for the Chinese government.

Fear of chaos and the value of social stability have a well-established history in Chinese cultural heritage, but the latter has taken on even more prominence since the rise of the discourse on building a "harmonious society" under the Hu-Wen leadership, which governed China between 2002 and 2012. A less euphemistic reading of this expression suggests little patience for contentious politics and activism more broadly. Indeed, as the cancer villages studied by Chen and his team and my three case studies suggest, a threat to social stability, real or perceived, seems to be the key factor precipitating government responses—though these responses vary widely. They may range from supportive to oppressive; local governments may intimidate villagers, but they may also carry out tests and regulate local industries more tightly. As much scholarship on China has highlighted (O'Brien and Li 2006; Cai 2010), the target to which complaints are addressed is a crucial factor influencing the response they may receive. When they are directed at local misimplementation, they are more likely to be met with a supportive response than if they take issue with the overall governing machinery of the Chinese state. When resistance unfolds, it is frequently (and strategically) couched in the language of "rightful resistance" (O'Brien and Li 2006)—relying on official laws and regulations—or of "rational resistance" (Johnson 2013a; A. Zhang 2014), which emphasizes scientific evidence, a solid knowledge base, and a balanced, reasonable approach. The implications of this point for China's environmental activism are significant. Given the emphasis on stability, environmentalism remains largely confined to embedded (non-oppositional) forms. A broader range for scaling up and reaching wider networks may be open in other areas of environmental action. Anti-dam campaigns, for instance, have shown signs of becoming more connected nationally and internationally (my personal communication with anti-dam campaigner, 2016). However, rural opposition to pollution has not so far become more interconnected (see also Fürst 2016).

When opposition to pollution is localized, individualized, often silenced, or seemingly absent, the assumption is that villagers must be either ignorant of the risks or unconcerned by them. One of my main aims in this book has been to show that this is not the case. Rural discontent about pollution is acute and widespread, even as much of it remains invisible. The relative invisibility of rural environmentalism could be interpreted in two ways. On the one hand, it means that much local discontent remains just that, local discontent. On the other hand, it means that citizens' anxieties over pollution are considerably broader and deeper than current reporting might suggest. This is only likely to increase as development pushes further and deeper into China's interior.

Chinese villagers are indeed aware of pollution and concerned about it, even if their concerns may not be manifested through collective contention. The question, then, is: Why not? This book has illustrated the complex dynamics that undermine the rise of collective environmental contention. Actions may not be simplistically mapped onto opportunities, nor are cost-benefit calculations straightforward. Definitions of what constitutes costs and benefits, and of what counts as successful activism, shift over time. The presence of collective action and the forms it may take are powerfully molded by expectations for a good life, a healthy body, and a clean environment, and by the ways in which toxic natures and toxicity of human bodies are normalized.

The analytical tool of resigned activism highlights that the most pervasive attitude to pollution among villagers is neither plain opposition nor complete complicity but rather ambivalence. This is the case even where locals seem to have accepted polluting activities in view of their economic benefits. This book shows that even those who depend on polluting activities for their livelihood resent pollution and are wary of how uneven the distribution of costs and benefits is. As I outlined in chapter 1, the Chinese government has increasingly taken action to improve environmental protection, including encouraging public participation. However, the deep embedding of polluting activities, not only in local economies, but also in locals' expectations poses obstacles to their potential role as vanguards of environmental regulation. Public participation is not a panacea. But citizens' roles as whistleblowers remain important in alerting higher authorities to pollution and putting pressure on them to intervene. The cases described here—where pollution becomes normalized and villagers no longer feel empowered to demand a cleaner environment—suggest that these internalized obstacles ought to be taken more seriously if the potential for citizen participation is to be realized more fully.

Like many scholars who work on activism and contentious politics in China, I am often asked whether the discontent I have witnessed, and the seeming rise in environmental protests, is a prelude to revolution or to democratization. My experiences and research in rural China have persuaded me to err on the side of caution. Radical change rarely happens suddenly. China's complex governance machine has developed very powerful ways of maintaining the current political order. Challenges to it are most often successfully metabolized to make the system even more resilient. There are of course critical events and junctures—such as major pollution incidents or the promulgation of new laws and regulations— which alter the status quo and may challenge people's ways of thinking and behaving. New technologies have undoubtedly furnished increasing portions of the population, including the rural population, with the ability to communicate efficiently, obtain information, and organize collective responses. But my fieldwork has led me to conclude that normalization is also an extremely powerful force, which decreases the likelihood of challenges to the status quo, particularly by those who already feel marginal and powerless.

This is not to say radical change is impossible. Of course, it's possible. But the challenges to it are complex and do not only originate from within the political system. We would ignore these subtler yet enduring obstacles at the peril of those who are most adversely affected by pollution and other forms of suffering. Without the benefit of having done in-depth research on more privileged social groups, I cannot conclude that resigned activism is less prevalent among these sections of the population, but I am inclined to guess that it characterizes in particular those at the relative margins of society and the economy. Indeed, it is deeply intertwined with marginalization.

In the specific shape it takes in this book, resigned activism may display some particular "Chinese characteristics." It is molded by the relatively limited space for political engagement and by a complex convergence of contexts. It also intersects with a culturally and historically rooted pride in the ability to endure hardship, often termed "eating bitterness." But this mixture of agency and acquiescence is by no means confined to China. Researching its prominence and the various forms it may take in other sociocultural, economic, and political contexts would yield extremely interesting and valuable comparisons. Indeed, as an analytical tool, resigned activism allows researchers to encompass empirical diversity and promotes conceptual innovation for the study of environmentalism and activism more broadly.

## Lessons for Studying Environmentalism and Global Environmental Justice

In all three cases studies examined here, pollution had deeply divisive effects. Pervasive processes of socioeconomic stratification that go hand in hand with development shape the ways in which the embedding of polluting activities affects local residents, both formally registered and migrants. Uneven entitlements and unequal access to capital—social, cultural, political, and economic—produce diverse expectations and create fragmented communities where individuals are less likely to form a sustained, collective front targeted at curbing pollution. At certain times, villagers negotiated directly with polluters, organized petitions, and mounted routinized blockades and small-scale protests. Yet, all of these forms of action remained rooted in the locality and their effects—in terms of curbing pollution—were limited. In Baocun, and, to a lesser extent, in Qiancun, these efforts were aimed at securing compensation (Baocun) or the right to mine (Qiancun), rather than at protecting the environment. The obstacles to environmental activism posed by community divisions play similarly important roles in areas affected by pollution beyond China, and require concerted scholarly attention.

A related and equally prominent effect of pollution on local socioeconomic structures is the complexity of the interface between the labels of victim and perpetrator. While in some cases one's identity as victim or perpetrator may seem relatively clear—by and large, poor migrants are victims and bosses of polluting businesses are perpetrators—identification is a much more intricate and uneven process. Regarding all those who live in the shadow of pollution as its victims would gloss over their multifaceted relationship to pollution and the uneven extent to which some draw economic benefits from polluting businesses. Conversely, regarding all those who are diversely connected to polluting activities as perpetrators—regardless of whether they manage polluting firms, are paid by them, or are involved in work that relies on a wider industrial ecology of pollution—also glosses over their uneven responsibility for pollution. This ambiguity significantly complicates attitudes toward pollution and affects the forms of activism embraced.

Self-identification raises further questions. Even poor migrants who, to a scholar of environmental injustice, may seem to irrefutably fit the category of victim may themselves repudiate this label, as Mr. Zhang's case in

Baocun illustrated. Individuals' subjectivity is complex and comprises the articulation of various, potentially conflicting identities and loyalties (Hall 1992). As a worker, Mr. Zhang was proud of his ability to earn a wage instead of being confined to depending on farming the land in his natal village. As a man, he was proud of the healthy, strong body that allowed him to take on strenuous and potentially harmful work. As a family man, he valued his ability to fulfill his moral obligations as a father and a son, and to provide for his parents and his children. Conversely, in Guiyu, e-waste workers who, upon a superficial analysis, may be categorized as perpetrators keenly protested their (relative) innocence by pointing out that their own work is not polluting, that local pollution has multiple sources beyond e-waste work, and that much of e-waste work is in fact environmentally friendly because it decreases waste. Their own assessment of their position most often conceded that some of e-waste work can be extremely hazardous, and that they are part of the wider system that sustains these harmful practices. But they were also at pains to highlight that they—and in particular their children—were exposed to these hazards, without much power to counter them. This illustrates the blurring of the categories of winners and losers.

Due to the complexity of subjectivity and the limited success of collective action, villagers also engaged with pollution on a more individual or family basis: purchasing bottled water, closing windows, wearing masks, avoiding the jobs deemed most harmful, and temporarily sending small children and pregnant women to live elsewhere. While cosmopolitan green campaigners may not regard some of these practices as activism, I argue that they nevertheless deserve attention as alternative, resigned forms of activism and environmental subjectivity, whereby those who live with pollution have become attuned to its presence, both as an environmental occurrence and a sociopolitical assemblage. Such engagements with the environment (natural, social, and political) alter what demands are regarded as acceptable and what strategies may be envisioned as feasible. They also mold different ways of valuing not only the environment but also health and development. Understanding resigned activism then requires a more encompassing, holistic, and diachronic study of pollution as it is experienced in its local contexts. Such attention will make visible the subtler and resigned forms of activism and challenge the existing parameters of what counts as activism.

Some scholars have illustrated worrying trends whereby engagement with environmental problems is shifting away from protecting the environment to protecting ourselves through lifestyle choices (Szasz 2007). These two forms of engagement need not be in conflict. Indeed, they could be mutually reinforcing. Resigned activism is not necessarily an obstacle to protecting the environment, somehow retreating into individualism. There are several layers in between, spanning from individuals to their families, communities, regions, nations, and the world as a whole. Citizens may move upward through these scales when opportunities arise and downward when their efforts for collective welfare prove less than satisfactory. But these shifts are neither stable nor straightforward, and they may give rise to conflicts and to entrenching pollution in poorer areas, especially when choices by certain individuals, communities, regions, or nations are made at the expense of others.

For this reason, environmental justice campaigners often approach NIMBY (not-in-my-back-yard) movements as selfish by virtue of their localistic focus; they are regarded only as embryonic forms of environmentalism that must escalate into broader advocacy movements in order to tackle the real origins of injustice and contamination. In China, however, at least at present, escalating localized complaints into broader social movements targeting the political, economic, and social origins of injustice is politically implausible. While environmental campaigners are skilled at navigating tumultuous political waters (Fürst 2016; Ho and Edmonds 2008; Mertha 2008; Yang 2010b; Yang and Calhoun 2007), the political space has undoubtedly narrowed in recent years. This point has broader implications for the study of global environmental justice. Rather than applying a single, universal parameter to assess the potential of localized forms of activism to tackle environmental problems effectively, local conditions must be taken seriously. What may be empowering in one sociopolitical and cultural context may be counterproductive elsewhere. Attention to resigned activism then is also intended to look beneath the surface and beyond the most recognizable forms of engagement with environmental injustice to the daily realities of entrenched pollution and what forms of agency they may engender and curtail.

In light of this, simply mandating stricter environmental regulations from the top down, or informing the public about the risks posed by pollution, does not necessarily ensure change at the local level. This does not

mean, however, that communities, campaigners, academics, and governments should stop trying to secure a cleaner environment. When opportunities arise, researchers should work more closely with local communities,
local environmental organizations, and NGOs to improve their capacity to
protect their environment and health, form networks with other affected
communities and campaigners, and share knowledge and strategies to
empower local populations to live in healthier environments.

One of the most basic principles of environmental justice is that the
people most affected must be able to speak for themselves and participate
fully in decision-making (First National People of Color Environmental
Leadership Summit, 1991). In order to do so, they must be endowed with
better tools, better networks, and more power to be heard. Access to and
understanding of technical and scientific evidence is a hurdle for citizens
and campaigners, and not only in China (Fürst 2016). If formal institutions
do little to assist them, alternatives should be mobilized. Researchers can
play an important role in these efforts, by collaborating with communities
and civil society organizations in gathering relevant evidence, and identifying tools to mobilize knowledge and gain recognition. A stronger sense of
their rights, better access to tools to measure their exposure to harm, and
broader networks with activists and campaigners who have faced similar
challenges elsewhere may allow members of affected communities to see
their suffering not as the inevitable result of circumstances, but as systemic
injustice that needs to be challenged. A clearer awareness of the impact that
the normalization of pollution and powerlessness among affected communities exerts upon their activism—as I pursued in this book—is a prerequi-
site for understanding how better to help them.

While I have argued against applying a one-size-fits-all parameter to
assess environmentalism in different local contexts, I strongly believe in
the value of demanding environmental justice for all. On this basis, should
we support global environmental standards for what acceptable levels of
contamination are, and expect and demand the same outcomes everywhere? The answer for me is a strong idealistic yes, but also a realistic sense
that the timelines for change will differ in different places and that challenges to implementation will be uneven. For instance, putting pressure on
governments overseeing communities exposed to lead contamination globally to comply to a global standard for what a safe level of lead is should
present a valuable tool for campaigners, though we may not expect all

localities will all comply immediately. If we as researchers are to be effective in our efforts, we must also learn—just as the members of affected communities have—the value of patience, perseverance, and of recognizing both opportunities and limitations, particularly when we work in authoritarian contexts.

Influential anthropologist Margaret Mead allegedly said: "Never doubt that a small group of thoughtful, committed citizens can change the world. Indeed, it is the only thing that ever has."[3] Sweeping change however rarely takes place overnight. Citizens' commitments may relent or shift; their allegiances may alter; their strategies and tactics may adapt to fluctuating contexts; and their expectations may transform. Attending to subtle forms of engagement, and to resignation as well as to activism, equips us better to grasp the circumstances in which change takes place, the many instances in which it does not, and the wide, perhaps most interesting, space between them. Emphasizing resigned activism may seem a pessimistic and gloomy move. My intention is quite the opposite: to restore attention to agency as it emerges in unlikely places and subtle forms. Recognizing the power of resignation and grasping its origins is the first step toward giving a voice to those whose experiences are otherwise doomed to remain invisible.

Last but not least, the environmental deterioration described in this book is not a uniquely Chinese problem. Nor does it only have Chinese origins. To the contrary, the world in which we live is deeply interconnected. The market for the commodities and materials produced and processed in Baocun, Qiancun, and Guiyu—fertilizers, heavy metals, mobile phones—is a global and uneven one. In such a profit-driven economic system, culpability for creating such deplorable conditions and unequal spread of benefits and harm goes well beyond China, and implicates us all. As is well documented, international political economy affects the uneven distribution of pollution (see for instance Pellow 2007; Walker 2012). China's recent role as the "world's factory" is part and parcel of such patterns of distribution. These local and global commodity chains have given rise to a vast number of "ecological distribution conflicts" (EJAtlas 2016) related to the unequal metabolism of the global economy (Martinez-Alier et al. 2016).

Responsibilities for these environmental injustices span from global governance and global capital to national governments, local jurisdictions, and down to each individual. Changing individual behavior, acting in more environmentally responsible ways—from buying fewer new technologies to eating less food produced with fertilizers and pesticides—may not solve these problems entirely, but neither should these efforts be dismissed outright. The Chinese slogan "everyone is responsible for protecting the environment" is not only empty propaganda or neoliberal-washing and delegating of responsibilities to individuals. We all can and should play our part. But this should go hand in hand with more environmentally responsible actions by national governments, markets, and global institutions.

A commitment to global environmental justice requires that, when environmental standards are applied to wealthier communities, regions, or nations, this should not result in pollution simply relocating among poorer and more marginalized ones, or, even worse, in deliberate plans to locate pollution among those deemed least able to oppose it. Conversely, such commitment also needs to include the recognition of existing inequalities in order to identify feasible and sustainable ways forward. Crackdowns, without careful consideration of how communities can live sustainably, are not solutions. Closing down mines or banning informal e-waste processing without a feasible plan for securing a livelihood for communities that rely on these activities are doomed to fail and to create further problems. A better understanding of the conditions under which pollution is perpetuated is a crucial step toward achieving a cleaner and healthier environment for all. By highlighting some of the complexities and challenges faced by those living with pollution in rural China, I hope this book is one small step in the right direction.

# Appendix: Methodological Strategies and Challenges

## Overview

The roots of my interests in pollution and my efforts at interdisciplinary collaboration lie in the deep frustration I experienced during my previous project on experiences of cancer in rural Langzhong, an area of northeast Sichuan province, which, as I discovered during fieldwork, had a relatively high incidence of stomach and esophageal cancers (Lora-Wainwright 2013a). Having watched villagers whom I had lived with for over a year die of cancer, I was haunted by the question of what caused higher rates of cancer in this area. My training in anthropology, however, did not equip me to answer it. The village was not industrialized, and neither was the county more broadly (though this has changed considerably now). There were no obvious factors (or clusters of factors) that would explain higher cancer incidence, or set this area apart from other neighboring localities with allegedly lower cancer rates. I set out to enlist the help of epidemiologists, but their responses were skeptical and emphasized the complexity of cancer causality, suggesting I should just focus on something else.

Through these exchanges, the question of evidence, or, more pertinently, of whose evidence counts, steadily climbed up my intellectual and civic list of concerns. This process persuaded me that cancer was not the 'right' problem to focus on. But it also convinced me of four additional things. First, the contested area of pollution's effects on health was a topic I wanted to research more in-depth. Second, I needed to team up with experts from other disciplines, particularly public health, epidemiology, and environmental studies, in order to do it justice. Third, I needed to choose sites where pollution was much more obvious. And fourth, I needed to shift the center of my attention from cancer alone to a flexible range of symptoms and illnesses that are broadly linked to the type of pollution affecting a given fieldsite.

With all of this in mind, since 2007 I invested several years establishing interdisciplinary collaborations with Chinese colleagues and identifying suitable fieldsites. As serendipity would have it, that same year the Social Science Research Council (SSRC) established the China Environment and Health Initiative (CEHI) led by Jennifer Holdaway, a program director at SSRC with extensive knowledge of China. CEHI organized a workshop in 2008 that I attended, and where I met several Chinese and international scholars, such as Benjamin Van Rooij, Wang Wuyi, and Li Liping who became essential contacts for establishing fieldwork in Baocun, Qiancun, and Guiyu, respectively. By 2009, CEHI had established strong contacts and collaborations with several Chinese institutions under the name FORHEAD (Forum for Health, Environment, and Development) and created a small grants program that funded, among other projects, research in Baocun. FORHEAD hosted its launch conference in Beijing in 2009, where I presented some of our initial analysis of Baocun and made more vital contacts for my future research, most notably Professor Chen Ajiang, who conducted pioneering research on China's "cancer villages." These networks, acquired through the vital resources provided by FORHEAD, form the essential backbone of this book.

In order to research pollution and health qualitatively and ethnographically, I employed a case study approach (see Small 2009). Extended periods of fieldwork were undertaken in three sites between 2009 and 2013 (see figure A.1). The first site is Baocun, a village in Yunnan province (southwest China) affected by phosphorous mining and processing, with several fertilizer plants situated close to village accommodation. These activities have affected the soil, water, and air, and may have caused several health effects, including fluorosis. The second is Qiancun, a village with a long history of lead and zinc mining in Hunan province (central China). Mining and mineral processing have severely affected agricultural livelihoods and caused a range of ailments. The third is Guiyu, a town comprising several villages that specialize in trading, dismantling, and processing electronic waste. Some of these processes leech heavy metals into the soil and water, cause serious air pollution, and have adverse effects on locals' health, particularly in the case of children.

While pollution differs across the three sites, their respective social, economic, and political contexts present productive points for comparison.

**Figure A.1**
Map of China and location of fieldsites.

The research sites were selected according to several related criteria. First, and perhaps most obvious, they needed to be sites where I could carry out qualitative social science research. On this front, my involvement in FOR-HEAD was essential in providing the networks needed to undertake this research. Given the sensitivity of the topic, I needed to select sites in collaboration with academic colleagues in China who may have carried out research there previously or who had strong networks with local governments. Second, I deliberately sought severely polluted areas, where the proximate relationship between pollution and illness would be relatively uncontested epidemiologically, even if causality and the extent of exposure might remain scientifically undetermined. This presented productive ground for interdisciplinary collaborations to study pollution, its effects, and villagers' responses to it from a range of perspectives.

As a consequence, I selected sites where natural scientists had already collected data on pollution, exposure, and health effects, or where I could team up with scientists interested in collecting this data. In each case, sites were selected through collaborations forged with the support of FORHEAD. FORHEAD provided financial support for fieldwork in Baocun, funded and coordinated research in Qiancun, and introduced me to Professor Li Liping, with whom I collaborated for fieldwork in Guiyu. Research in the first two sites was part of two separate but related interdisciplinary projects on environmental health threats, involving teams composed almost exclusively of Chinese researchers. Research in the third site, Guiyu, was part of a different type of collaboration, with a fellow anthropologist who carried out comparative research in Japan (Peter Wynn Kirby) and with Professor Li Liping, a leading public health scholar at Shantou University, who previously undertook research in Guiyu and whose contacts were instrumental in securing access to the site. As several public health studies of the environmental health effects of local pollution already existed for this site, our research focused on the social science angle. These case studies were complemented by a review of groundbreaking work by Professor Chen Ajiang and members of his research team, who have carried out qualitative research in some of China's "cancer villages" for the past decade (see chapter 2), much of which was also funded by FORHEAD. As much as I wished to undertake fieldwork in locations that had been identified by locals and/or by the media as "cancer villages," it was never feasible for me to do so. I am therefore extremely grateful to Professor Chen for allowing me to summarize and analyze some of his case studies here.

As with any project, the resulting book is not the book I had intended to write. I hoped to produce a fully rounded ethnographic account of pollution and health in rural China, which included the perspectives of local officials and industry bosses. However, the extremely sensitive nature of the research topic and the fragility of some local collaborations set their own limits, and I opted instead to focus most closely on local communities. More often than not, as the book illustrates, local firms and the local government are deeply intertwined, making open questioning on pollution with both these social groups detrimental to future attempts to access the fieldsites. In Qiancun's case, doing interviews with these particular actors was another team member's responsibility. In Baocun, village officials and factory bosses were suspicious, if not openly obstructive. Probing too much into local officials' and local bosses' perspectives may have compromised my ability to carry out fieldwork among villagers in the way I envisioned.

Although being part of larger projects was a great learning opportunity, it also brought its own difficulties. In Qiancun, for instance, a team of the medical geographers (most prominently Wang Wuyi, Yang Linsheng, and Li Yonghua), who had originally carried out research in the area with the help of the county Center for Disease Control (CDC), established access to the village. The leaders of the county CDC, Teng and Peng, were immensely supportive and participated in some project activities, which included attending numerous meetings and working on public education materials. While the CDC granted unique access at certain times, we also had to accept that at other times our presence was inconvenient and politically infeasible given the sensitivity of heavy metals pollution. Indeed, CDC leaders would be the first to be placed under scrutiny if the project caused tensions locally. Being a foreigner made my research visits even more sensitive than those by Chinese colleagues, and it eventually became politically impossible for me to carry out further fieldwork there at all.

Villagers themselves showed signs of research exhaustion, having been the targets of repeated rounds of medical and (later) social science research, and not all were welcoming of more research visits. Conversely, attempts to minimize this problem by carrying out fieldwork in conjunction with other team members was hindered by different timetables, which rarely allowed researchers to visit the site at the same time. Beyond the practical challenges lay a set of conceptual ones. Different disciplines speak different languages, have different ideas of what counts as data, and what methods are appropriate to gather it. These differences took time to manage by finding the right synergy and mutual respect among team members. But they also required compromises on the part of team members. In my case, the length of time spent in the field had to be reduced, requiring more reliance on key informants and on interviews. Respecting the delicate balance of teamwork also means there are challenges and tensions that it would be unethical to discuss here, and which must remain within the confines of team meetings.

## Methodology Employed by Chen and His Team in "Cancer Villages"

Due to the extreme political sensitivity of the topic of "cancer villages," I was unable to carry out fieldwork in these sites firsthand. However, I believe that a review of existing research on some of these sites is extremely valuable as a comparative canvas for my own sites. Chapter 2 of this book draws on research carried out by China's most prominent environmental sociologist working on "cancer villages," professor Chen Ajiang. Chen and his team published several articles and co-authored a book published in 2013, all of which draw on fieldwork in several cancer villages between 2005 and 2011. This research was supported by China's National Social Science Funding Program and by FORHEAD. Chen and his team have since undertaken follow-up fieldwork on some of these sites, also funded by FORHEAD, but my account in this book only covers their findings until 2011. The cases examined in chapter 2 include: Shangba village in northern Guangdong province; Xiqiao village, and particularly the subvillage of Dingbang, in northern Zhejiang; Jian'nan village in southeast Jiangxi; a village known as Likeng village due to its proximity to the homonymous incinerator in Guangdong province; Dongjing village in northern Jiangsu; and Huangmengying in the Huai River Basin, Hebei province.

Researchers focused fieldwork investigations around the broad questions of rural environmental change and the relationship between environment and health. The research focus evolved gradually and became more refined through increasingly close engagements with each site. Research methods were mostly qualitative, relying on observation, semi-structured and unstructured interviews, and informal conversations. In some villages, questionnaires were also employed. Interview subjects included villagers, village doctors, members of the village elite (such as well-educated and well-connected villagers), industry managers, and representatives of the village committees and of relevant government bureaus at the township and county level. Researchers closely examined the historical development of any polluting firms (some of which had since closed), the manufacturing processes they employed, and the pollution they may have produced.

During the process of data collection, a wealth of documentary sources were consulted, in particular those provided by villagers, including images, maps, and a variety of documents. Most notably, these included name lists of cancer deaths compiled by villagers and documentary evidence prepared for lawsuits by villagers. Chen and his team also consulted official medical records, results of relevant water tests, and investigations carried out by local Centers for Disease Control and Environmental Protection Bureaus, to better understand local levels of pollution and disease incidence. Rather than taking any of these records at face value, Chen and his team painstakingly crosschecked the accuracy of evidence wherever possible, and strived to determine the extent to which a relationship between cancer and pollution was present, and what other factors played a role in the onset of cancer. Chen's training in both natural and social sciences endowed him with a rare ability to understand the issues at stake and interpret the evidence. This multidimensional methodology proved extremely valuable and allowed Chen's team to differentiate between cases where the evidence for pollution causing cancer was relatively strong and those where it was weak or nonexistent (Holdaway and Wang 2013).

## Doing Fieldwork

Given that fieldwork in Baocun and Qiancun followed a shared model, and was in both cases part of larger collaborative projects, I discuss these first, and then turn to fieldwork in Guiyu separately.

Baocun was studied in conjunction with the NGO Yunnan Health and Development Research Association (YHDRA), as part of the China Environment and Health Initiative (CEHI) supported by the Social Science Research Council under the aegis of FORHEAD. The members of the research team were myself (representing anthropology), two scholars of politics and law, a public health scholar, two sociologists, and a law student. YHDRA recruited a small team of researchers, including one of its own project officers, Zhang Yiyun. I served as a key consultant to the project and also as a fieldworker. Fieldwork was arranged with the assistance of two key consultants, Wang Qiliang from Yunnan University's School of Law and Benjamin Van Rooij (now at University of California Irvine)—and all members of the research team (except Van Rooij and Wang) carried out fieldwork in conjunction. Van Rooij played a crucial role in providing the contacts necessary to enable fieldwork, based on his long-term research at the site. He also provided a wealth of background information on Baocun, which formed an excellent basis for subsequent data collection and analysis (see Van Rooij 2006).

Qiancun was also studied under the aegis of CEHI and FORHEAD, as a collaboration between SSRC and China's Institute of Geographical Sciences and Natural Resources Research (IGSNRR). IGSNRR researchers had already carried out pilot studies on levels of lead in the soil, in the rice, and in the hair and blood of local residents in the area for roughly a decade. This provided some valuable, if limited, data on the effects of mining on local environment and health in the present. A previous grant from SSRC had enabled IGSNRR to examine the pathways of exposure to heavy metals and to begin work on developing public education materials. In 2009, social scientists, including myself, formed a FORHEAD multidisciplinary team alongside IGSNRR researchers to examine the social, economic, and political context in which mining is situated.

All data for Baocun and Qiancun was collected between April 2009 and September 2012, over several research visits. Fieldwork was conducted in five phases: April–June 2009 (Baocun phase 1); July–August 2009 (Baocun phase 2); August 2010 (Qiancun phase 1); September 2011 (Baocun phase 3); and August–September 2012 (Qiancun phase 2). Based on anthropological and sociological research protocols, we conducted participant observation and semi-structured interviews with a total of 160 respondents. Fieldworkers lived with local families during research, which provided them with vital opportunities for more informal exchanges, including observing and participating in a range of activities, such as helping with farming and household chores. Reflecting on these activities, as the anthropologist, I took daily notes on local living conditions, local diet, and informal conversations (that were not audio-recorded). In both sites, semi-structured interviews were undertaken in pairs with a lead researcher accompanied by a research assistant. They lasted a minimum of half an hour and in some cases several hours. Some villagers were interviewed twice to clarify particular points raised in the first interview. A sub-sample of research participants whose views differed from the majority was also re-interviewed to maximize the diversity of responses. Interviews were audio-recorded and subsequently transcribed in full in Chinese.

In Qiancun, after two brief exploratory visits by the full team in 2009 and 2010, team members divided the project into separate (though interdependent) sections, and each took charge of their sub-projects. Due to different academic schedules and competing commitments among team members, scholars carried out data collection at different times, and reported on their key findings to other team members after completing data collection. Data collected during early visits by one set of scholars helped those who were planning later visits to revise and fine-tune their questions. I carried out fieldwork in 2010 (together with sociologist Dr. Lu Jingfang) and in 2012 (with two research students, coordinated in conjuction with Dr. Lu Jixia), doing a total of over sixty in-depth interviews with villagers, including two village doctors.

Interviewees were recruited by snowball sampling from an initial opportunistic sample beginning with the local host family and their most immediate neighbors. The participation rate was extremely high, with only a very limited number of villagers declining to be interviewed, and almost always because they were occupied with farming or household tasks at the time. In both sites, interviewees included a broad range of local residents including men and women, younger and older, formal residents and migrant workers, those employed by industries and mines, and those engaged in other activities, such as farming or running small shops. This approach was intended to include perspectives from those differently positioned vis-à-vis industry and mining, and who are therefore likely to benefit from these activities in unequal ways. As snowballing may introduce a bias in data collection and affect conclusions drawn, we made efforts to identify different (conflicting) social groups in each village and to interview members of all these groups. Spending extended periods of time in both sites and listening to conversations among locals allowed us to map the different factions of each local community. Carrying out research in several phases also allowed us to identify any bias or gaps in the interview sample and address it during the next phase.

Research in Guiyu demanded a rather different strategy. The sensitive nature of this fieldsite as a major hub for e-waste trade and processing and a media hotspot made long-term participant observation infeasible. The needed permissions to settle in the area would likely never be granted and given the detrimental effects of foreign media attention on the local economy, locals would at best be unwelcoming toward a foreign resident, at worst openly hostile. My local host and collaborator, Professor Li Liping (Shantou University) was inimitable in devising alternative strategies to conduct in-depth qualitative research. She introduced me to Juanjuan, a university student whose hometown is close to Guiyu. Through Juanjuan's exceptional kindness, I was able to stay with her family for brief, but extremely intensive, visits in September 2012 and June 2013, and meet her relatives and friends who are involved in the e-waste trade and processing. My colleague Peter Wynn Kirby and my student Loretta Lou also took part in some of the fieldwork. In the course of these visits I conducted participant observation and ten in-depth interviews, as well as extensive ethnographic interviews with Juanjuan, her family members and her close neighbors. A few families, including Juanjuan's, have become friends and precious key informants who welcomed me into their e-waste workshop and patiently shared their experience with me over several insightful conversations. However, language was another obstacle to full-blown participant observation. Though I speak Mandarin, the local dialect is radically different, making it impossible to follow conversations among locals that did not involve me directly.

To compensate for these shortfalls, and with the help of Professor Li, I recruited ten students at Shantou University whose hometowns are close to Guiyu and who speak the local dialect. I trained them in social science research methods over several days, explaining my research focus, discussing the interview outline, and conducting in-depth structured interviews with them so they could learn through practice. After an initial trial period, five of these students carried out forty in-depth interviews in Guiyu, which they recorded and transcribed in Mandarin in full. As several of the interviewees were familiar with the interviewer, the tone of interviews is usually informal and relaxed, resembling closely the kind of exchange typical of anthropological fieldwork. Given the limited amount of participant observation possible, this seemed the best approximation and the most effective way of obtaining an 'insider's view' on local experiences.

As in Baocun and Qiancun, research participants were recruited by snowball sampling from an initial opportunistic sample beginning with families of the interviewees' former classmates and friends. To reduce bias and cover a wide range of local residents, research assistants were required to include residents differently positioned vis-à-vis e-waste processing. Interviewees included some village-level officials, a family who runs a private quality certification office, traders, managers, and workers (including migrant workers) of small e-waste businesses, ranging from plastic processing to capacitor recycling and pc board smelting. Research participants also included villagers not directly involved in the industry. Though attempts were made to interview evenly across gender lines, men tended to be more outspoken and willing to be interviewed; therefore, unfortunately, women are underrepresented. Dr Luo Yajuan assisted me in analyzing interview transcripts and locating relevant media reports, policy documents, and official statements, which provided an invaluable counterpart to interviews and fieldwork.

## Interview Design and Data Analysis

An initial topic guide was inductively generated based on my years of experience living in the Chinese countryside and understanding of rural life, and through extensive reading on China's rural industrialization, as well as the comparative literature about environmental health justice movements. The first interview guideline was developed in March 2009 in preparation for fieldwork in Baocun, within the overarching framework of FORHEAD and the small grants program. Each team member drafted a set of questions that would ensure coverage across an interdisciplinary set of concerns. The team considered them over several meetings until consensus was reached on a question list. This process also ensured that fieldworkers were clear about the rationale for each question and for their order. The same process was followed for Qiancun and for Guiyu. Interview guidelines, and the area of inquiry more broadly, were routinely adjusted as part of the research process. During each stage of data collection, researchers compared initial findings and modified areas of enquiry accordingly. Further fine-tuning took place following the completion of each stage of research and before commencing subsequent stages. Given the hermeneutic process of constantly refining questions throughout data collection, the processes of data collection and analysis became to a large extent inseparable. This is one of anthropology's trademarks. There are of course drawbacks to frequently revising interview guidelines and adapting them to different sites, to the extent that such variations make data less easy to measure or analyze schematically. However, this approach is more sensitive to how locals themselves frame problems, and more suited to qualitatively research a politically and emotionally sensitive topic that research participants may not be willing to speak about at first.

Since research in all three sites forms part of my larger and long-term overarching endeavor to understand Chinese villagers' experiences and responses to pollution and related health problems, the questioning approach was the same across all of them. In Baocun and Guiyu, I oversaw the design of the entire project's research questions, and the overall focus was on villagers' experiences and responses to pollution. In Qiancun, by contrast, I was responsible for a subset of questions that dealt with this focus in particular, while other team members explored other angles and employed other approaches and methodologies. My own focus on villagers' experiences is best suited to my training and overlaps substantially with research questions explored in Baocun and Guiyu, therefore making the data sets comparable. Questions and comments found to be particularly effective at eliciting answers in Baocun were initially employed in Qiancun and Guiyu, and gradually tuned to local specificity. While the overall topics remained the same, more specific questions were developed based on responses gathered in each site and tailored to their particular local history, as well as the social, political, and economic context. By raising questions about specific villagers or key local events, interviewers also showed familiarity with the local setting, which increased the research participants' trust toward researchers and their willingness to discuss sensitive topics. This in turn substantially improved the quality of the data collected.

All three projects were presented to the local community in very broad terms, as studies of local livelihood and health. No mention was made of pollution. Partly, this was intended to ease access to the sites. While researching attitudes to health is potentially sensitive, it is much less sensitive than researching views on pollution's health effects in an area where pollution is plain for all to see, and industry and resource processing are so central to local revenue. This strategy for framing the projects was also of vital importance to data collection. Introducing the research as focused on the link between pollution and health would have sensitized respondents to this concern and probably biased answers to questions about pollution's effects on health toward stressing such effects. This would not have allowed researchers to assess how serious concerns about pollution's effects on health were for the local population, and how they relate to other concerns.

With this in mind, fieldworkers spent the first part of the interview asking questions about the individual's health, recent or recurring illness episodes, and how they explain them. This was intended to gauge whether pollution was raised as one of the potential causes of illness and how prominently it figures in their experience. We presented our aim as understanding local life conditions, or health, and made open comments to life narratives—for instance "your life is hard"—in order to elicit reactions. Comments such as this were identified in previous interviews as reflections frequently used by villagers to summarize their experiences. Employing such comments in interviews was a way of resorting to a culturally sensitive discourse that was rooted locally, to express sympathy for the research participant, and to increase their trust in the researcher. We also shared our experiences since moving to the site (some fieldworkers suffered with sore throat in Baocun, and I had a nosebleed the night I arrived there) as a way of eliciting comments and establishing a relaxed atmosphere of mutual exchange.

If research participants did not raise the subject of pollution, researchers asked general questions about the environment, to test whether this elicited any mention of pollution, and potentially of its effects on health. Almost every interviewee raised the subject of pollution at this stage, and many reflected on its potential health effects. When our research participants mentioned pollution, we asked a very broad question: "What to do?" (*zenme ban*). In this way, we gathered a range of responses without suggesting what we envisioned to be likely pathways of action. As a whole, this interview structure allowed us to assemble a picture of how villagers conceived of the relationship between illness and pollution, and how it was situated within their everyday life and experience more broadly.

The opportunity to live with local families in Baocun and Qiancun considerably improved our ability to establish informal relationships with locals. For instance, in Baocun I allowed the subvillage head to film me while I helped villagers carry bags of grain, and he aired the footage on his daily village TV program that evening. The following day, villagers were much more welcoming and readily invited me into their homes. Bringing a cake to a child's birthday party at a migrant's home (which included many other migrants) similarly opened up access to several local families. At the end of each interview in all three sites, we offered research participants small gifts, such as hand towels, washing powder, or fruit, as a small token of appreciation, which significantly improved our relationships with locals.

In-depth analysis of interview transcripts and fieldnotes through several rounds of coding highlighted some key recurrent concepts through which villagers made sense of their experiences. The most prominent were uneven distribution of benefits and power, uncertainty, and inevitability. Having identified these concepts as central, I returned to the data collected to examine how these concepts were presented and discussed, and how they came into being. On the basis of interviews, I compiled a timeline of key events and phases of development in all three sites, which allowed me to map: changes in livelihood strategies and income from mining, industry, and e-waste processing; the changing texture of the local communities; social relations between villagers; and the shifting relationships among villagers, local officials, and polluting firms. I then examined how the core concepts identified through analysis of interviews and fieldnotes mapped onto these changes, narratives, and pathways of action.

One of the main immediate findings was that in severely polluted areas such as these fieldsites, pollution and its health effects have become so routinized that they hardly seem worth mentioning during interviews. Often villagers would initially reply to questions on their health that it was "good"; they neglected to mention common problems like nose infections or hand and feet swelling because they had already become normalized. As the discussion progressed, it became clear that villagers who defined themselves as healthy were in fact affected by several more or less acute illnesses. It also became clear that these conditions were seen as a normal part of life. In this context, the task of the anthropologist is to raise such questions as: What is normal? What is seen to be "good health"? And why? How has this changed over time? To put it most broadly, how do people come to think what they think and do what they do? What must their local worlds be like for them to think and act the way they do? These are some of the questions this book set out to answer.

# Notes

## Introduction

1. All place names below the county, with the exception of Guiyu, are pseudonyms. As was the case for Guiyu, the relatively high profile of some of the sites discussed in chapter 2 and reference to media reports made it impossible to disguise their identity. All individuals' names have been changed to protect the identity of research participants.

2. I follow the Chinese custom of referring to people first by their surname and second by their name throughout the book.

3. As I explain in chapter 4, Qiancun was an "administrative village" composed of four subvillages. Administrative villages (*xingzheng cun*) are the smallest bureaucratic entity in China's complex administrative hierarchy. An administrative village may overlap with a preexisting "natural village" (*ziran cun*), whose residents sometimes share the same surname and can be traced to a single ancestor, or it may incorporate several natural villages.

4. Despite Bourdieu's aim of overcoming the dichotomy between structure and agency through these concepts, he has often been criticized for replicating the structural determinism he set out to avoid. I will not rehearse these discussions here. For some examples, see: Calhoun 1993; Fowler 1997; Jenkins 1992; and Swartz 1997.

## Chapter 1

1. Although the elite versus non-elite dichotomy is not useful in understanding the complexities of local interests and stratification (see Gaventa 1980, 130–131), the question remains important.

2. See also Pierre Bourdieu (1977, 1990) for a sociological explanation of how people learn to view change as impossible and deny themselves opportunities they do not feel entitled to. This point is elaborated further in the introduction.

3. A critical review of the wide range of literature on political ecology is beyond the scope of this chapter. Some important works in this area include: Forsyth 2003; Latour 2004; Peet and Watts 2004; and Robbins 2004.

4. This has become a vast field. For some of the most important contributions, see: Allen 2003; Boudia and Jas 2014; Brown 2007; Brown, Morello-Frosch, and Zavestoski 2012; Callon, Lascoumes, and Barthe 2009; Edelstein 2004; Frickel et al. 2010; Hess 2007; Ottinger and Cohen 2011; Fortun 2001; Jasanoff 2012; Michaels 2008; Proctor 1995; and Proctor and Schiebinger 2008.

5. Studies of citizen science include: Allen 2003; Brown 2007; Callon, Lascoumes, and Barthe 2009; Corburn 2005; Irwin and Wynne 1996; and Leach, Scoones, and Wynne 2005.

6. For an earlier elaboration of what, following Gaventa 1980, we called "activist acquiescence," see Van Rooij, Lora-Wainwright, Wu, and Zhang 2014. In this book, the focus is further refined to examine the connotations and implications of resigned activism beyond access to justice (which was the focus of the earlier paper), to reflect more on the mutual shaping between perceptions and practices and to broaden the definition of activism.

7. See, for instance: Allen 2003; Auyero and Swistun 2009; Brown and Mikkelsen 1997; Checker 2005; Fortun 2001; Lerner 2005; Petryna 2002; Shevory 2007; and Smartt Gullion 2015.

8. Recent scholarship has pointed out that the high turnover of cadres at the local level may hinder state-led greening growth initiatives by producing short-time horizons and a focus on visible results rather than long-term efficiency and effectiveness. Cadres' limited knowledge of the area may also contribute to unrealistic targets (Eaton and Kostka 2014).

9. It is also possible that coverage of such events has increased, rather than their incidence per se.

10. In employing the term environmental consciousness, I do not wish to imply this is a measurable entity, nor that some may have 'more' of it than others. It is intended to point to the diverse ways in which people relate to the environment.

## Chapter 2

1. A shorter and earlier version of this chapter appeared as: "China's cancer villages: contested evidence and the politics of pollution," by Anna Lora-Wainwright and Ajiang Chen, in *A Companion to the Anthropology of Environmental Health*, ed. M. Singer (Chichester, UK; Malden, MA: John Wiley & Sons, Inc., 2016), 396–416. I am grateful to Wiley Blackwell for permission to revise and publish this version of the chapter.

2. While the term "cancer village" remains a contentious and contested one, granting the use of single quotation marks, employing them throughout the chapter seemed rather awkward, and I have therefore opted to omit them.

3. Confusingly, it seems that the database was revised between July and November 2013, so the search on November 10, 2013 only returned sixty-eight reports, not 106 as in the original July search.

4. A new groundbreaking atlas of cancer and pollution in the Huai River Basin is a notable exception (Yang and Zhuang 2013), though its data is aggregated at the county, rather than village, level.

5. For an excellent English language overview, see L. Liu (2010). See Deng (2009) for an influential report in in the Hong Kong-based magazine *Phoenix Weekly*. See also Lora-Wainwright and Chen (2016) on the question of scientific evidence.

6. FORHEAD is a collaboration between the Social Science Research Council and several Chinese partner institutions including most prominently China's Institute of Geographical Sciences and Natural Resources Research. See the appendix.

7. Due to the relatively high profile of many of these sites, anonymizing proved impossible for some locations. However, names of individuals have been changed, as is customary, to protect their identity.

8. This account is a summary of findings discussed by Chen and Cheng (2011) and Li and Cheng (2013).

9. The value of Chinese yuan has fluctuated considerably over the past decade. At its lowest, one GBP was equivalent to fifteen yuan, at its highest one GBP was equivalent to ten yuan.

10. By contrast, a neighboring village that also lacked clean water, but where villagers did not share the same surname and lacked organizational capacity was unable to obtain clean water.

11. This account is a summary of findings discussed by Cheng (2013).

12. This account is a summary of findings discussed by Cheng and Li (2013) and Chen and Cheng (2011).

13. This account is a summary of findings discussed by Li and Chen (2013); see also Johnson (2013a) and Zhang (2014).

14. This account is a summary of findings discussed by Luo (2012, 2013).

15. Barefoot lawyer is the literal translation of the Chinese term used to describe a villager who trained him or herself in law.

16. "According to Chinese law, in compensation lawsuits concerning environmental pollution, the polluter carries the burden of proof with respect to ...

demonstrating the lack of causal link between the polluter's actions and the harmful result. In practice, however, courts are still known to require plaintiffs to produce evidence sufficient to demonstrate causation. Given the difficulty of proving causation in environmental pollution cases, this reversal of burden of proof is often the critical determinant of outcome in environmental litigations" (Wang 2006, 209).

17. This account is a summary of findings discussed by Cheng (2013).

## Chapter 3

1. The names of the villages and of all research participants are pseudonyms.

2. For more details see the appendix.

3. In 2009, one British pound corresponded to roughly thirteen yuan, though exchange rates between 2009 and 2013 fluctuated between ten and fifteen yuan per pound.

4. On the process of symbiosis between industries, local residents, and officials, see Lora-Wainwright et al. 2012; Van Rooij 2006; and Van Rooij et al. 2012, 2014.

5. For a more detailed account of these processes, see Lora-Wainwright et al. 2012; and Van Rooij et al. 2012.

6. This process of normalization and its implications are described more fully in Lora-Wainwright 2013d.

## Chapter 4

1. All names of people and places, except for Fenghuang, are pseudonyms.

2. According to "The Land Administration Law of the People's Republic of China," urban land is owned by the state, and rural land is owned by collectives (villages or subvillages) (the Central People's Government of the People's Republic of China 2005).

3. According to the revised national standard in 2008 (which is not officially promulgated yet), the acceptable lead content in soil was lowered to 80μg/g, just a third of the previous standard.

4. Water in category V is deemed unsuitable for any use, too toxic even to touch.

5. It is important to note that other factors that contribute to human capital and even health—such as housing conditions, access to education, better overall nutrition—may have improved due to the additional income from mining. I am grateful to Jennifer Holdaway for pointing this out.

6. The extent to which pollution decreased since the closing of the mines remains unclear. Tests in 2013 (N. Chen 2013) showed lower levels of pollution than those carried out by IGSNRR in previous years (Y. Li 2012); however, the different location of samples may affect the results. Indeed, the latest tests were more broadly scattered, whereas those by IGSNRR were more concentrated close to tailings and watercourses.

7. For more information on the project see FORHEAD (2013). A key requirement of FORHEAD funding is that projects should be problem-focused and action-oriented, intended to provide the knowledge base for better responses, including better policy or local solutions. As its co-directors suggest, FORHEAD supports combining natural and social sciences to gather data to inform interventions that may target environmental health problems effectively, be understood and supported by local people, and enable a better integration of interventions aimed at environmental health impacts with development strategies (Holdaway and Wang 2013).

## Chapter 5

1. Due to the highly sensitive nature of this fieldsite, my Chinese collaborator, Professor Li Liping, initially stipulated that a pseudonym should be employed. As a consequence, in previous publications I referred to the site as Treasure Town (Kirby and Lora-Wainwright 2015b; Lora-Wainwright 2016). It became clear, however, that the town was easily identifiable given its peculiar position, and we have therefore opted to use its real name. To protect the anonymity of interviewees, all their names have been changed, as is common practice. For more details on the methodology employed and access issues, in particular Professor Li's incredible and creative support, refer to the appendix.

2. There is no space here to list the numerous reports on Guiyu. Recent examples include: CBS 2008; the Ecologist 2010a, 2010b; Moskvitch 2012; and Watson 2013.

3. Predictably, these are no longer available online. See Lora-Wainwright (2016) on some locals' reactions in the immediate aftermath of the reports.

4. For instance, based on local interviews, Minter (2013) highlighted that the majority of the profits in Guiyu are drawn from reuse rather than from recovery of materials. In different ways, Gabrys (2011), Lepawsky (2015b), and Lepawsky and Mather (2011) have highlighted that the entire process whereby appliances are discarded and recycled is much more complex than the snapshot view we might glean from only looking at end-of-life stages (where components no longer work) or from focusing only on a subset of processes in places like Guiyu.

5. Lepawsky, Goldstein, and Schulz (2015) have shown that estimates reproduced in reports by UNEP and Greenpeace (Baldé et al. 2015; Brigden et al. 2005; Rucevska

et al. 2015; Wang et al. 2013) are often self-referential and based on weak evidence and that press releases tend to highlight the worst case scenarios.

6. Lepawsky (2015a, 2015b) demonstrated the recent rise of a "post-Basel world," where much of the e-waste trade takes places within the developing world rather than from developed to developing nations. For some efforts to rethink e-waste flows see Kirby and Lora-Wainwright (2015a).

7. Elsewhere I have shown that e-waste workers stress the important and sustainable contribution of the informal sector and are skeptical of formalization (Lora-Wainwright 2016). The Chinese government's condemnation of informal recycling intentionally ignores that it leads to a higher rate of recovery of reusable materials (Lepawsky, Goldstein, and Schulz 2015; Minter 2013). This point is also ignored by the latest UNEP report (Rucevska et al. 2015), which focuses instead on the most dangerous and harmful subsections of informal recycling (material recovery) in order to support formalization and technologization as a solution. The feasibility of such plans should also be under scrutiny. While the UNEP report claims that informal recycling has been phased out in areas where it formerly thrived, journalistic investigations show that it simply pushed informal recycling into rural areas (Lepawsky, Goldstein, and Schulz 2015). This mirrors my findings in Guiyu that banned activities are still carried out, but further from the spotlight.

8. Since my last visit in 2013, some progress was made with the circular economy park. My doctoral student, Carlo Inverardi Ferri, reported that during his visit to Guiyu in 2014 the park was under construction and that some workshops had begun to move in. The television manufacturer TCL also seemed to be a major feature of the park, processing old cathode ray tube televisions.

## Conclusion

1. Since 2014 there are signs that attempts to formalize and regulate the sector may have met with some success (Carlo Ferri and Yvan Schulz, personal communication, 2015), but only more detailed fieldwork could reveal the precise dynamics and their effects, and whether this change is temporary.

2. It is of course hard to determine whether environmental contention itself has increased, or whether it has simply gained more coverage in the media and more attention by various levels of the Chinese government.

3. Although this quote is widely attributed to Mead, its accuracy remains disputed.

# Bibliography

Agamben, Giorgio. *State of Exception*. Translated by K. Attell. Chicago, IL: University of Chicago Press, 2005.

Agamben, Giorgio. *Homo Sacer: Sovereign Power and Bare Life*. Translated by D. Heller-Roazen. Redwood City, CA: Stanford University Press, 1998.

Agrawal, Arun. *Environmentality: Technologies of Government and the Making of Subjects*. Durham, NC: Duke University Press, 2005.

Ahlers, Anna L., and Gunter Schubert. "Strategic Modeling: 'Building a New Socialist Countryside' in Three Chinese Counties." *China Quarterly* 216 (2013): 831–849.

Ahlers, Anna L., and Gunter Schubert. "Building a New Socialist Countryside—Only a Political Slogan?" *Journal of Current Chinese Affairs* 38, no. 4 (2009): 35–62.

Alexander, C., and J. Reno, eds. *Economies of Recycling: The Global Transformation of Materials, Values and Social Relations*. London: Zed Books, 2012.

Allen, Barbara. *Uneasy Alchemy: Citizens and Experts in Louisiana's Chemical Corridor Disputes*. Cambridge, MA: MIT Press, 2003.

Ansfield, Jonathan. "Alchemy of a Protest: The Case of Xiamen PX." In *China and the Environment: The Green Revolution*, edited by Sam Geall, 136–201. London: Zed Books, 2013.

Auyero, Javier, and Debora Swistun. *Flammable: Environmental Suffering in an Argentine Shantytown*. Oxford: Oxford University Press, 2009.

Baike.com. "Huai River Warriors NGO." Accessed January 20, 2016. http://www.baike.com/wiki/淮河卫士民间环保组织.

Baldé, C. P., F. Wang, R. Kuehr, and J. Huisman. *The Global E-waste Monitor – 2014: Quantities, Flows and Resources*. Bonn: United Nations University, Institute for the Advanced Study of Sustainability–SCYCLE, 2015.

Bales, Kevin. *Disposable People: New Slavery in the Global Economy*. Berkeley: University of California Press, 1999. (Third edition, updated with a new preface, 2012.)

Balshem, Martha. *Cancer in the Community: Class and Medical Authority.* Washington, DC: Smithsonian Institution Press, 1993.

Basel Action Network (BAN). *Exporting Harm: The High-Tech Trashing of Asia.* Seattle: Basel Action Network, 2002. Accessed September 26, 2016. http://svtc.org/wp-content/uploads/technotrash.pdf.

*BBC News Online.* "China Acknowledges 'Cancer Villages,'" February 22, 2013. Accessed February 22, 2013. http://www.bbc.co.uk/news/world-asia-china-21545868.

Bebbington, Anthony, Denise Humphreys Bebbington, Jeffrey Bury, Jeannet Lingan, Juan Pablo Muñoz, and Martin Scurrah. "Mining and Social Movements: Struggles Over Livelihood and Rural Territorial Development in the Andes." *World Development* 36, no. 12 (2008): 2888–2905.

Bebbington, Anthony, Leonith Hinojosa, Denise Humphreys Bebbington, Maria Luisa Burneo, and Ximena Warnaars. "Contention and Ambiguity: Mining and the Possibilities of Development." *Development and Change* 39, no. 6 (2008): 887–914.

Boudia, S., and N. Jas, eds. *Powerless Science? Science and Politics in a Toxic World.* London: Berghahn, 2014.

Bourdieu, Pierre. *Language and Symbolic Power.* Cambridge: Polity Press, 1991.

Bourdieu, Pierre. *The Logic of Practice.* Cambridge: Polity Press, 1990.

Bourdieu, Pierre. *Distinction: A Social Critique of the Judgement of Taste.* London: Routledge, 1984.

Bourdieu, Pierre. *Outline of a Theory of Practice.* Cambridge: Cambridge University Press, 1977.

Boyd, Olivia. "The Birth of Chinese Environmentalism: Key Campaigns." In *China and the Environment: The Green Revolution*, edited by Sam Geall, 40–94. London: Zed Books, 2013.

Bullard, R. D. *Dumping in Dixie: Race, Class, and Environmental Quality.* 3rd ed. Boulder, CO: Westview Press, 2000.

Branigan, Tania. "China Targets 7.5% Growth and Declares War on Pollution," *The Guardian*, March 5, 2014. Accessed March 6, 2014. http://www.theguardian.com/world/2014/mar/05/china-pollution-economic-reform-growth-target.

Brettell, Anna. "The Politics of Public Participation and the Emergence of Environmental Proto-Movements in China." PhD diss., University of Maryland, 2003.

Bridge, Gavin. "Contested Terrain: Mining and the Environment." *Annual Review of Environment and Resources* 29, no. 1 (2004): 205–259.

Brigden, Kevin, Iryna Labusnka, David Santillo, and Michelle Allsopp. "Recycling of Electronic Wastes in China and India: Workplace and Environmental

Contamination–Report." Greenpeace International, 2005. Accessed January 20, 2016. http://www.greenpeace.org/international/PageFiles/25134/recycling-of-electronic -waste.pdf.

Brown, Philip M. *Toxic Exposures: Contested Illnesses and the Environmental Health Movement.* New York: Columbia University Press, 2007.

Brown, Philip M. "Popular Epidemiology and Toxic Waste Contamination: Lay and Professional Ways of Knowing." In *Illness and the Environment: A Reader in Contested Medicine,* edited by Philip M. Brown, Steve Kroll-Smith, and Valerie Gunter, 364–383. New York: New York University Press, 2000.

Brown, Philip M., and Edwin J. Mikkelsen. *No Safe Place: Toxic Waste, Leukemia, and Community Action.* Berkeley: University of California Press, 1997.

Brown, Philip M., R. Morello-Frosch, and S. Zavestoski, eds. *Contested Illnesses: Citizens, Science, and Health Social Movements.* Berkeley: University of California Press, 2012.

Brown, Philip M., Steve Kroll-Smith, and Valerie Gunter. "Knowledge, Citizens and Organizations: An Overview of Environments, Diseases, and Social Conflicts." In *Illness and the Environment: A Reader in Contested Medicine,* edited by Philip M. Brown, Steve Kroll-Smith, and Valerie Gunter, 9–25. New York: New York University Press, 2000.

Cai, Yongshun. *Collective Resistance in China: Why Popular Protests Succeed or Fail.* Stanford, CA: Stanford University Press, 2010.

Cai, Yongshun. "Collective Ownership or Cadres-Ownership? The Non-Agricultural Use of Farmland in China." *China Quarterly* 175 (2003): 662–680.

Calhoun, Craig. *Bourdieu: Critical Perspectives.* Cambridge: Polity Press, 1993.

Callon, Michel, Pierre Lascoumes, and Yannick Barthe. *Acting in an Uncertain World.* Cambridge, MA: MIT Press, 2009.

Cao, Lin. "'Cancer Village' Label Is Opposed by Villagers," *Xinhua Meiri Dianxun,* April 11, 2013.

Cao, Wulin. "Shangba Villagers Drink Safe Water," *Shaoguan Ribao,* February 21, 2009.

Carmin, J. A., and J. Agyeman, eds. *Environmental Inequalities Beyond Borders: Local Perspectives on Global Inequities.* Cambridge, MA: MIT Press, 2011.

Carson, Rachel. *Silent Spring.* London: Penguin, 2000.

Carter, N., and A. P. J. Mol, eds. *Environmental Governance in China.* London: Routledge, 2007.

*CBS News*. "Following the Trail of Toxic E-Waste," November 9, 2008. Accessed January 20, 2016. http://www.cbsnews.com/news/following-the-trail-of-toxic-e-waste/.

CCTV (China Central Television). "Rivers and Villages," 2004. Accessed March 4, 2005. http://www/cctv/com/news/china/20040810/102281.html. (No longer online.)

Chan, Anita. *China's Workers Under Assault. The Exploitation of Labor in a Globalizing Economy*. London: M. E. Sharpe, 2001.

Central People's Government of the People's Republic of China. *The Land Administration Law of the People's Republic of China*, 2005. Accessed July 11, 2013. http://www.gov.cn/banshi/2005-05/26/content_989.htm.

Centers for Disease Control and Prevention. "Low Level Lead Exposure Harms Children: A Renewed Call for Primary Prevention," January 2012. Accessed March 16, 2017. https://www.cdc.gov/nceh/lead/acclpp/final_document_030712.pdf.

Center for Legal Assistance to Pollution Victims (CLAPV). "Environmental Public Interest Litigation Case," June 13, 2012. Accessed January 20, 2016. http://www.clapv.org/english_lvshi/zhichianjian_content.asp?id=59.

Chan, Anita, and Kaxton Siu. "Chinese Migrant Workers: Factors Constraining the Emergence." In *China's Peasants and Workers: Changing Class Identities*, edited by Beatriz Carrillo and David S. G. Goodman, 79–101. Cheltenham, UK: Edward Elgar, 2012.

Chan, Anita, Richard Madsen, and Jonathan Unger. *Chen Village: Revolution to Globalization*. 3rd ed. Berkeley: University of California Press, 2009.

Chan, Chris King-Chi. *The Challenge of Labour in China: Strikes and the Changing Labour Regime in Global Factories*. London: Routledge, 2010.

Chan, Chris King-Chi, and Pun Ngai. "The Making of a New Working Class? A Study of Collective Actions of Migrant Workers in South China." *China Quarterly* 198 (2009): 287–303.

Chan, Jenny. "A Suicide Survivor: The Life of a Chinese Worker." *New Technology, Work, and Employment* 28, no. 2 (2013): 84–99.

Chan, Jenny, and Ngai Pun. "Suicide as Protest for the New Generation of Chinese Migrant Workers: Foxconn, Global Capital, and the State." *The Asia-Pacific Journal* 8, issue 37, no. 2 (2010): 1–33.

Chan, Jenny, Ngai Pun, and Mark Selden. "The Politics of Global Production: Apple, Foxconn, and China's New Working Class." *The Asia-Pacific Journal* 11, issue 32, no. 2 (2013): 1–22.

Chan, Jenny, and Mark Selden. "China's Rural Migrant Workers, the State, and Labor Politics." *Critical Asian Studies* 46, no. 4 (2014): 599–620.

Checker, Melissa. *Polluted Promises: Environmental Racism and the Search for Justice in a Southern Town.* New York: New York University Press, 2005.

Chen, Aimin, and Jie Gao. "Urbanization in China and the Coordinated Development Model—The Case of Chengdu." *Social Science Journal* 48 (2011): 500–513.

Chen, Ajiang. "Inward and Outward Perspectives on 'Cancer Villages.'" *Journal of Guangxi University for Nationalities Philosophy and Social Science Edition* 3, no. 2 (2013): 68–74. Revised and included in *Cancer Village Research: Understanding and Responding to Environmental Health Risks*, edited by Ajiang Chen, Pengli Cheng, Yajuan Luo, Caihong Li, and Qi Li, 1–28. Beijing: Chinese Social Science Press, 2013.

Chen, Ajiang. *Secondary Anxiety: A Social Interpretation of Water Pollution in the Tai Lake Basin.* Beijing: China Social Sciences Press, 2010.

Chen, Ajiang, and Pengli Cheng. "Understandings and Responses to the Risk of 'Cancer-Pollution'" *Xuehai* 3 (2011): 30–41. Revised and included in *Cancer Village Research: Understanding and Responding to Environmental Health Risks*, edited by Ajiang Chen, Pengli Cheng, Yajuan Luo, Caihong Li, and Qi Li, 171–198. Beijing: Chinese Social Science Press, 2013.

Chen, Ajaing, Pengli Cheng, Yajuan Luo, Caihong Li, and Qi Li, eds. *Cancer Village Research: Understanding and Responding to Environmental Health Risks.* Beijing: Chinese Social Science Press, 2013.

Chen, Nengchang. "Heavy Metal Pollution and Food Safety." Paper presented at the SSRC FORHEAD Annual Conference, Beijing, China, November 2013.

Cheng, Pengli. "A Rich 'Cancer Village.'" In *Cancer Village Research: Understanding and Responding to Environmental Health Risks*, edited by Ajiang Chen, Pengli Cheng, Yajuan Luo, Caihong Li, and Qi Li, 67–97. Beijing: Chinese Social Science Press, 2013.

Cheng, Pengli, and Caihong Li. "The Coexistence of Poverty and Cancer." In *Cancer Village Research: Understanding and Responding to Environmental Health Risks*, edited by Ajiang Chen, Pengli Cheng, Yajuan Luo, Caihong Li, and Qi Li, 98–120. Beijing: Chinese Social Science Press, 2013.

Chi, Xinwen, Martin Streicher-Porte, Mark Y. L. Wang, and Markus A. Reuter. "Informal Electronic Waste Recycling: A Sector Review with Special Focus on China." *Waste Management (New York, N.Y.)* 31, no. 4 (2011): 731–742.

*China Daily.* "'Cancer Village' in Spotlight," May 10, 2004. Accessed September 26, 2016. http://www.chinadaily.com.cn/english/doc/2004-05/10/content_329171.htm.

China Environment Net. "Provincial Party Committee Secretary Comments Twice on the Effects of Multidepartmental Strike Action on Guiyu's E-waste Pollution Staged Clean-up." Published electronically December 5, 2012. Accessed September

9, 2013. http://www.cenews.com.cn/xwzx/zhxw/ybyw/201212/t20121204_733157
.html.

China News Week. "Likeng, We Are Sorry for You," vol. 2009044, December 7, 2009.
Accessed September 26, 2016. http://magazine.sina.com/bg/chinanewsweek/2009/
2009044/index.html.

Clapp, Jennifer. Toxic Exports: The Transfer of Hazardous Wastes from Rich to Poor
Countries. Ithaca, NY: Cornell University Press, 2001.

Cody, Edward. "Chinese Newspapers Put Spotlight on Polluters," The Washington
Post, May 25, 2004. Accessed May 26, 2004. http://www.washingtonpost.com/
wp-dyn/articles/A53012-2004May24.html.

Corburn, Jason. Street Science: Community Knowledge and Environmental Health Justice.
Cambridge, MA: MIT Press, 2005.

Das, Veena. "Suffering, Legitimacy and Healing. The Bhopal Case." In Illness and the
Environment: A Reader in Contested Medicine, edited by Philip M. Brown, Steve Kroll-
Smith, and Valerie Gunter, 270–286. New York: New York University Press, 2000.

Davis, Deborah S. "Demographic Challenges for a Rising China." Daedalus: The
Journal of the American Academy of Arts and Sciences 143, no. 2 (2014): 26–38.

De Certeau, Michel. The Practice of Everyday Life. Berkeley: University of California
Press, 1984.

Deng, Fei. "China's 100 Cancer Causing Places," Phoenix Weekly no. 11, April 2009.

Deng, Jia. "Cancer Villages on the Banks of the Shaying River," China Society News,
May 19, 2005.

Deng, Yanhua, and Guobin Yang. "Pollution and Protest in China: Environmental
Mobilization in Context." Special issue, China Quarterly 214 (2013): 321–336.

Dewey, Caitlin. "Chinese State Media Release a Map Showing the Spread of 'Cancer
Villages,'" Washington Post, February 22, 2013. Accessed February 23, 2013. https://
www.washingtonpost.com/news/worldviews/wp/2013/02/22/chinese-state-media
-releases-a-map-showing-the-spread-of-cancer-villages/.

Diamant, N. J., S. B. Lubman, and K. J. O'Brien, eds. Engaging the Law in China: State,
Society, and Possibilities for Justice. Stanford, CA: Stanford University Press, 2005.

Double Leaf. "China Cancer Villages Map." Created May 7, 2009. Updated June 1,
2009. Accessed November 12, 2009. https://www.google.com/maps/d/viewer?mid
=zITChDulPwt4.k2riwzSMvTHk&hl=en&ie=UTF8&lr=lang_en&msa=0&ll=34.09872
8%2C117.292099&spn=0.268943%2C0.4422&z=11.

Eaton, Sarah, and Genia Kostka. "Authoritarian Environmentalism Undermined?
Local Leaders' Time Horizons and Environmental Policy Implementation in China."
China Quarterly 218 (2014): 359–380.

*Ecologist.* "Low-Cost E-Waste Recycling in China Releasing Catalogue of Pollutants." September 3, 2010 (2010a). Accessed January 20, 2016. http://www.theecologist.org/ News/news_round_up/582564/lowcost_ewaste_recycling_in_china_releasing _catalogue_of_pollutants.html.

*Ecologist.* "UN Warns India and China over Growing Problem of E-Waste." February 22, 2010 (2010b). Accessed January 20, 2016. http://www.theecologist.org/News/ news_round_up/420967/un_warns_india_and_china_over_growing_problem_of _ewaste.html.

Economy, Elizabeth. *The River Runs Black: The Environmental Challenge to China's Future.* Ithaca: Cornell University Press, 2004.

Edelstein, Michael. *Contaminated Communities: Coping with Residential Toxic Exposure.* Boulder, CO: Westview, 2004. First published 1988.

Edin, Maria. "State Capacity and Local Agent Control in China: CCP Cadre Management from a Township Perspective." *China Quarterly* 173 (2003): 35–52.

*EJAtlas. Environmental Justice Atlas.* 2016. Accessed June 30, 2016. http://ejatlas.org/.

Elvin, Mark. *The Retreat of the Elephants: An Environmental History of China.* New Haven, CT: Yale University Press, 2004.

Erickson, Kai. *Everything in Its Path: Destruction of Community in the Buffalo Creek Flood.* New York: Simon and Schuster, 1976.

Farmer, Paul. *Pathologies of Power: Health, Human Rights, and the New War on the Poor.* Berkeley: University of California Press, 2003.

Fedorenko, Irina, and Yixian Sun. "Microblogging-Based Civic Participation on Environment in China: A Case Study of the PM 2.5 Campaign." *VOLUNTAS: International Journal of Voluntary and Nonprofit Organizations* 26, no. 4 (2015): 1–29.

First National People of Color Environmental Leadership Summit. *17 Principles of Environmental Justice.* Adopted October 27, 1991, Washington, DC. Accessed July 6, 2016. http://www.ejnet.org/ej/principles.html.

FORHEAD (Forum on Health, Environment, and Development). *Rural Mining and Public Health: A Multi-disciplinary Study in Hunan Province (2010-present).* September 17, 2013. Accessed September 26, 2016. http://www.forhead.org/en/support/cqyj/ hn-fh/.

Forsyth, Tim. *Critical Political Ecology: The Politics of Environmental Science.* London: Routledge, 2003.

Fortun, Kim. *Advocacy After Bhopal: Environmentalism, Disaster, New Global Orders.* Chicago, IL: Chicago University Press, 2001.

Fowler, Bridget. *Pierre Bourdieu and Cultural Theory: Critical Investigations*. London: Sage, 1997.

Fraser, Nancy. *Justice Interruptus: Critical Reflections on the "Postsocialist" Condition*. New York: Routledge, 1996.

Frickel, S., and K. Moore, eds. *The New Political Sociology of Science: Institutions, Networks, and Power*. Madison, WI: University of Wisconsin Press, 2006.

Frickel, Scott, Sahra Gibbon, Jeff Howard, Joanna Kempner, Gwen Ottinger, and David J. Hess. "Undone Science: Charting Social Movement and Civil Society Challenges to Research Agenda Setting." *Science, Technology & Human Values* 35 (2010): 444–473.

Fürst, Kathinka. "Regulating Through Leverage: Civil Regulation in China." Unpublished PhD diss., University of Amsterdam, 2016.

Fürst, Kathinka. "The Rise of Non-Governmental Organizations as Regulators of Industrial Pollution in China." Paper presented at "Implementation of Environmental Law in China." Beijing, China, August 23, 2012.

Gabrys, Jennifer. *Digital Rubbish: A Natural History of Electronics*. Ann Arbor, MI: University of Michigan Press, 2011.

Gallagher, Mary. "China's Workers Movement and the End of the Rapid-Growth Era." *Daedalus: The Journal of the American Academy of Arts and Sciences* 143, no. 2 (2014): 81–95.

Galtung, Johan. "Violence, Peace, and Peace Research." *Journal of Peace Research* 6, no. 3 (1969): 167–191.

Gaventa, John. *Power and Powerlessness: Quiescence and Rebellion in an Appalachian Valley*. Oxford: Clarendon Press, 1980.

Geall, Sam, ed. *China and the Environment: The Green Revolution*. London: Zed Books, 2013.

Geall, Sam. "Interpreting Ecological Civilisation, Part Two." *Chinadialogue*, July 8, 2015. Accessed January 20, 2016. https://www.chinadialogue.net/article/show/single/en/8027-Interpreting-ecological-civilisation-part-two.

Gottlieb, Robert. *Environmentalism Unbound: Exploring New Pathways for Change*. Cambridge, MA: MIT Press, 2002.

Gould, Kenneth A., Allan Schnaiberg, and Adam S. Weinberg. *Local Environmental Struggles: Citizen Activism in the Treadmill of Production*. Cambridge: Cambridge University Press, 1996.

Gramsci, Antonio. *Selections from the Prison Notebooks*. New York: International Publishers, 1971.

Gu, Hongyan. "NIMBYism in China: Issues and Prospects of Public Participation in Facility Siting." *Land Use Policy* 52 (2016): 527–534.

Guha, Ramchandra. *The Unquiet Woods: Ecological Change and Peasant Resistance in the Himalaya*. Berkeley: University of California Press, 2000. First published 1989.

Guha, Ramachandra, and Joan Martínez-Alier. *Varieties of Environmentalism: Essays North and South*. London: Earthscan, 1997.

Gunson, A. J., and Yue Jian. "Artisanal Mining in the People's Republic of China: Minerals, Mining, and Sustainable Development." Paper published by the International Institute for Environment and Development (IIED), no. 74, London, September 2001. http://pubs.iied.org/pdfs/G00719.pdf.

Guo, Xiaolin. "Land Expropriation and Rural Conflicts in China." *China Quarterly* 166 (2001): 422–439.

Hall, Stuart. "The Question of Cultural Identity." In *Modernity and its Futures*, edited by Stuart Hall, David Held and Anthony McGrew, 274–316. Cambridge: Polity Press in association with the Open University, 1992.

Hardt, Michael. "Affective Labor." *Boundary 2* 26, no. 2 (1999): 89–100.

Hathaway, Michael. *Environmental Winds: Making the Global in Southwest China*. Berkeley: University of California Press, 2013.

He, Guangwei. "Special Report: the Legacy of Hunan's Polluted Soil." *Chinadialogue*, July 7, 2014. Accessed January 20, 2016. https://www.chinadialogue.net/article/show/single/en/7076-Special-report-the-legacy-of-Hunan-s-polluted-soils.

Hess, David. *Alternative Pathways in Science and Industry: Activism, Innovation, and the Environment in an Era of Globalization*. Cambridge, MA: MIT Press, 2007.

Hildebrandt, Timothy. *Social Organizations and the Authoritarian State in China*. Cambridge: Cambridge University Press, 2013.

Ho, Peter. "Introduction: Embedded Activism and Political Change in a Semi-Authoritarian Context." In *China's Embedded Activism: Opportunities and Constraints of a Social Movement*, edited by Peter Ho and Richard L. Edmonds, 1–19. London, New York: Routledge, 2008.

Ho, P., and R. L. Edmonds, eds. *China's Embedded Activism: Opportunities and Constraints of a Social Movement*. London: Routledge, 2008.

Ho, Samuel P. S. *Rural China in Transition: Non-Agricultural Development in Rural Jiangsu, 1978–1990*. Oxford: Clarendon Press, 1994.

Hofrichter, R., ed. *Reclaiming the Environmental Debate: The Politics of Health in a Toxic Culture*. Cambridge, MA: MIT Press, 2000.

Holdaway, Jennifer. *Environment, Health and Migration: Towards a More Integrated Analysis.* Geneva: United Nations Research Institute for Social Development, 2014. http://www.unrisd.org/holdaway.

Holdaway, Jennifer. "Environment and Health Research in China: The State of the Field." Special issue, *China Quarterly* 214 (2013): 255–282.

Holdaway, Jennifer. "Environment and Health in China: An Introduction to an Emerging Research Field." Special issue, *Journal of Contemporary China* 19, no. 63 (2010): 1–22.

Holdaway, Jennifer, and Wuyi Wang. English preface. In *Cancer Village Research: Understanding and Responding to Environmental Health Risks,* edited by Ajiang Chen, Pengli Cheng, Yajuan Luo, Caihong Li, and Qi Li, 14–24. Beijing: Chinese Social Science Press, 2013.

Holdaway, Jennifer, and Lewis Husain. *Food Safety in China: A Mapping of Problems, Governance, and Research.* Beijing: FORHEAD (Forum on Health, Environment, and Development). February 2014. Accessed September 26, 2016. http://www.forhead.org/uploads/soft/140306/1-140306103309.pdf.

Holifield, Ryan, Michael Porter, and Gordon Walker. *Spaces of Environmental Justice.* Oxford: Wiley, 2010.

Hook, Leslie. "China Smog Cuts 5.5 Years from Average Life Expectancy," *Financial Times,* July 8, 2013. Accessed January 21, 2016. http://www.ft.com/cms/s/0/eed7c0be-e7ca-11e2-9aad-00144feabdc0.html.

Hook, Leslie. "Chinese Riot Police Clash with Protesters," *Financial Times,* July 3, 2012. Accessed January 21, 2016. http://www.ft.com/cms/s/0/60ad9fae-c4dd-11e1-b6fd-00144feabdc0.html.

Horowitz, Leah. "Micropolitical Ecology: Power, Profit, Protest: Grassroots Resistance to Industry in the Global North." *Capitalism, Nature, Socialism* 23, no. 3 (2012): 21–34.

Hu, Tao, and Bryan Tilt. "Environmental Impact Assessment in China: Current Procedures and Evaluation of Case Studies." Paper presented at "Implementation of Environmental Law in China, Beijing, China," August 23, 2012.

Huo, Xia, Lin Peng, Xijin Xu, Liangkai Zheng, Bo Qiu, Zongli Qi, Bao Zhang, Dai Han, and Zhongxian Piao. "Elevated Blood Lead Levels of Children in Guiyu, an Electronic Waste Recycling Town in China." *Environmental Health Perspectives* 115, no. 7 (2007): 1113–1117.

Hurst, William. *The Chinese Worker after Socialism.* Cambridge: Cambridge University Press, 2009.

IPE: Institute of Public and Environmental Affairs. Accessed May 12, 2014. www.ipe.org.cn.

Irwin, Alan, and Brian Wynne. *Misunderstanding Science? The Public Reconstruction of Science and Technology*. Cambridge: Cambridge University Press, 1996.

IT Engineers for Environmental Protection Association. "Danger Maps" (*weixian ditu*). Accessed January 20, 2016. http://z.epmap.org/ngo.

Jasanoff, Sheila. *Science and Public Reason*. London: Routledge, 2012.

Jasper, James. *The Art of Moral Protest: Culture, Biography and Creativity in Social Movements*. Chicago, IL: Chicago University Press, 1997.

Jenkins, Richard. *Pierre Bourdieu*. London: Routledge, 1992.

Johnson, Thomas. "The Health Factor in Anti-Waste Incinerator Campaigns in Beijing and Guangzhou." Special issue, *China Quarterly* 214 (2013a): 356–375.

Johnson, Thomas. "The Politics of Waste Incineration in Beijing: The Limits of a Top-Down Approach?" Special issue, *Journal of Environmental Policy and Planning* 15, no. 1 (2013b): 109–128.

Johnson, Thomas. "Environmentalism and NIMBYism in China: Promoting a Rules-Based Approach to Public Participation." *Environmental Politics* 19, no. 3 (2010): 430–448.

Kanbur, R., and X. Zhang, eds. *Governing Rapid Growth in China: Equity and Institutions*. London: Routledge, 2009.

Kasperson, Roger E., and Jeanne X. Kasperson. "Hidden Hazards." In *The Social Contours of Risk: Volume I: Publics, Risk Communication and the Social Amplification of Risk*, edited by Roger E. Kasperson and Jeanne X. Kasperson, 115–132. London: Earthscan, 2005.

Kirby, Peter Wynn. *Troubled Natures: Waste, Environment, Japan*. Honolulu, HI: University of Hawaii Press, 2011.

Kirby, Peter Wynn, and Anna Lora-Wainwright, eds. *Peering Through Loopholes, Tracing Conversions: Remapping the Transborder Trade in Electronic Waste*. Special section, *Area* 47, no. 1 (2015a): 4–47.

Kirby, Peter Wynn, and Anna Lora-Wainwright. "Exporting Harm, Scavenging Value: Transnational Circuits of E-Waste between Japan, China and Beyond." In Kirby and Lora-Wainwright, *Peering Through Loopholes, Tracing Conversions*, Special section, *Area* 47, no. 1 (2015b): 40–47.

Kirsch, Stuart. "Indigenous Movements and the Risks of Counterglobalization: Tracking the Campaign against Papua New Guinea's Ok Tedi Mine." *American Ethnologist* 34, no. 2 (2007): 303–321.

Klawiter, Maren. *The Biopolitics of Breast Cancer: Changing Cultures of Disease and Activism*. Minneapolis, MN: University of Minnesota Press, 2008.

Kleinman, Arthur. *Social Origins of Distress and Disease: Depression, Neurasthenia, and Pain in Modern China*. New Haven, CT: Yale University Press, 1986.

Kostka, Genia, and Arthur P. J. Mol. "Implementation and Participation in China's Local Environmental Politics: Challenges and Innovations." Special issue, *Journal of Environmental Policy and Planning* 15, no. 1 (2013): 3–16.

Kroll-Smith, S., P. M. Brown, and V. Gunter, eds. *Illness and the Environment: A Reader in Contested Medicine*. New York: New York University Press, 2000.

Latour, Bruno. *Politics of Nature: How to Bring the Social Sciences into Democracy*. Cambridge, MA: Harvard University Press, 2004.

Lazzarato, Maurizio. *Lavoro Immateriale: Forme di Vita e Produzione di Soggettività*. Verona: Ombre Corte, 1996.

Leach, Melissa, and Ian Scoones. "Mobilising Citizens: Social Movements and the Politics of Knowledge." IDS Working Paper 276, Institute of Development Studies, Brighton, UK (2007): 1–34. Accessed September 26, 2016. http://www.ids.ac.uk/files/Wp276.pdf.

Leach, Melissa, Ian Scoones, and Andy Stirling. *Dynamic Sustainabilities: Technology, Environment, Social Justice*. London: Earthscan, 2010.

Leach, M., I. Scoones, and B. Wynne, eds. *Science and Citizens: Globalization and the Challenge of Engagement*. London: Zed Press, 2005.

Lee, Ching Kwan. "State and Social Protest." *Daedalus: The Journal of the American Academy of Arts and Sciences* 143, no. 2 (2014): 124–134.

Lee, Ching Kwan. "Pathways of Labor Activism." In *Chinese Society: Change, Conflict and Resistance*, 3rd ed., edited by Elizabeth J. Perry and Mark Selden, 57–79. London: Routledge, 2010.

Lee, Ching Kwan. *Against the Law: Labor Protests in China's Rustbelt and Sunbelt*. Berkeley: University of California Press, 2007.

Lee, Ching Kwan, and Yuan Shen. "China: The Paradox and Possibility of a Public Sociology of Labor." *Work and Occupations* 36, no. 2 (2009): 110–125.

Lee, Kingsyhon, and Ming-Sho Ho. "The Maoming Anti-PX Protest of 2014." *China Perspectives* 3 (2014): 33–39.

Lepawsky, Josh, and Charles Mather. "From Beginnings and Endings to Boundaries and Edges: Rethinking Circulation and Exchange through Electronic Waste." *Area* 43, no. 3 (2011): 242–249.

Lepawsky, Josh, and Mostaem Billah. "Making Chains That (Un)make Things: Waste-Value Relations and the Bangladeshi Rubbish Electronics Industry." *Geografiska Annaler, Series B, Human Geography* 93, no. 2 (2011): 121–139.

Lepawsky, Josh, Joshua Goldstein, and Yvan Schulz. "Criminal Negligence?" *Discard Studies: Social Studies of Waste, Pollution, and Externalities*, June 24, 2015. Accessed January 20, 2016. http://discardstudies.com/2015/06/24/criminal-negligence/.

Lepawsky, Josh. "Are We Living in a Post-Basel World?" *Area* 47 (1) (2015a): 7–15.

Lepawsky, Josh. "The Changing Geography of Global Trade in Electronic Discards: Time to Rethink the E-Waste Problem." *Geographical Journal* 181, no. 2 (2015b): 147–159.

Lerner, Steve. *Sacrifice Zones: The Front Lines of Toxic Chemical Exposure in the United States*. Cambridge, MA: MIT Press, 2010.

Lerner, Steve. *Diamond: A Struggle for Environmental Justice in Louisiana's Chemical Corridor*. Cambridge, MA: MIT Press, 2005.

Li, Caihong, and Pengli Cheng. "Problematisation and Stigmatisation." In *Cancer Village Research: Understanding and Responding to Environmental Health Risks*, edited by Ajiang Chen, Pengli Cheng, Yajuan Luo, Caihong Li, and Qi Li, 121–141. Beijing: Chinese Social Science Press, 2013.

Li, Gao, Shen Lei, and Yi Wang. "Regional Effects of the Exploitation of Mineral Resources." *Resources & Industries* 14, no. 6 (2012): 88–92.

Li, Junde. "Huai River Pollution Causes Several 'Cancer Villages,'" *Xinhua Meiri Dianxun,* June 29, 2004.

Li, Keqiang. "The Development of China's Mining Industry Makes an Important Positive Contribution to Global Economic Recovery," *China's mining industry resource website*, November 17, 2010. Accessed September 11, 2012. Available from http://www.kyzyw.com.cn/news/shownews.asp?id=615.

Li, Qi, and Ajiang Chen. "Fabricated Data on 'High Lung Cancer Incidence': A Sociological Investigation." *Nanjing Industrial University Journal Social Science Edition* 12, no. 1 (2013): 117–127. Revised and published as "Behind 'High Cancer Incidence.'" In *Cancer Village Research: Understanding and Responding to Environmental Health Risks*, edited by Ajiang Chen, Pengli Cheng, Yajuan Luo, Caihong Li, and Qi Li, 142–170. Beijing: Chinese Social Science Press, 2013.

Li, Y. "Environmental Health Risks." Paper presented at the SSRC FORHEAD Annual Conference, Beijing, China, November 7–9, 2012.

Li, Y., Y. Ji, L. Yang, and S. Li. "Effects of Mining Activity on Heavy Metals in Surface Water in Lead-Zinc Deposit Area." *Journal of Agro-Environment Science* 26, no. 1 (2007): 103–107.

Li, Y., W. Wang, L. Yang, and H. Li. "Environmental Quality of Soil Polluted by Mercury and Lead in Polymetallic Deposit Areas of Western Hunan Province." *Environmental Sciences* 20, no. 3 (2005): 187–191.

Li, Y., X. Zhang, L. Yang, and H. Li. "Levels of Cd, Pb, As, Hg, and Se in Hair of Residents Living in Villages around Fenghuang Polymetallic Mine, Southwestern China." *Bulletin of Environmental Contamination and Toxicology* 89 (2012): 125–128.

Li, Zhiyuan, Zongwei Ma, Tsering Jan van der Kujip, Zengwei Yuan, and Lei Huang. "A Review of Soil Heavy Metal Pollution from Mines in China: Pollution and Health Risk Assessment." *Science of the Total Environment* 468–469 (2014): 843–853.

Little, Peter. *Toxic Town. IBM, Pollution and Industrial Risks.* New York: New York University Press, 2014.

Litzinger, Ralph. "Search of the Grassroots: Hydroelectric Politics in Northwest Yunnan." In *Grassroots Political Reform in Contemporary China*, edited by Elizabeth J. Perry and Merle Goldman, 282–300. Cambridge, MA: Harvard University Press, 2007.

Liu, Lee. "Made in China: Cancer Villages," *Environment Magazine*, March/April 2010. Accessed May 1, 2010. http://www.environmentmagazine.org/Archives/Back%20Issues/March-April%202010/made-in-china-full.html.

Lock, Margaret. *Encounters with Aging: Mythologies of Menopause in Japan and North America.* Berkeley: University of California Press, 1993.

Lora-Wainwright, Anna. "The Trouble of Connection: E-Waste in China between State Regulation, Development Regimes and Global Capitalism." In *The Anthropology of Postindustrialism: Ethnographies of Disconnection*, edited by Ismael Vaccaro, Krista Harper, and Seth Murray, 113–131. London: Routledge, 2016.

Lora-Wainwright, Anna. *Fighting for Breath: Living Morally and Dying of Cancer.* Honolulu: University of Hawaii Press, 2013a.

Lora-Wainwright, Anna. "Dying for Development: Illness and the Limits of Citizens' Agency in China." Introduction to Special Collection, *China Quarterly* 214 (2013b): 243–254.

Lora-Wainwright, Anna. "Plural Forms of Evidence and Uncertainty in Environmental Health: A Comparison of Two Chinese Cases." *Evidence & Policy* 9, no. 1 (2013c): 49–64.

Lora-Wainwright, Anna. "'The Inadequate Life:' Rural Industrial Pollution and Lay Epidemiology in China." Special issue, *China Quarterly* 214 (2013d): 302–320.

Lora-Wainwright, Anna. "Of Farming Chemicals and Cancer Deaths: The Politics of Health in Contemporary Rural China." *Social Anthropology* 17, no. 1 (2009): 56–73.

Lora-Wainwright, Anna, and Chen Ajiang. "China's Cancer Villages: Contested Evidence and the Politics of Pollution." In *A Companion to the Anthropology of Environmental Health*, edited by Merrill Singer, 396–416. Malden, MA: John Wiley and Sons, 2016.

Lora-Wainwright, Anna, Yiyun Zhang, Yunmei Wu, and Benjamin Van Rooij. "Learning to Live with Pollution: The Making of Environmental Subjects in a Chinese Industrialized Village." *China Journal (Canberra, A.C.T.)* 68 (2012): 106–124.

Lukes, Steven. *Power. A Radical View*. London: Macmillan, 1974.

Luo, Yajuan. "'Cancer Village' in Subei." In *Cancer Village Research: Understanding and Responding to Environmental Health Risks*, edited by Ajiang Chen, Pengli Cheng, Yajuan Luo, Caihong Li, and Qi Li, 29–66. Beijing: Chinese Social Science Press, 2013.

Luo, Yajuan. "Environmental Struggle among Rural Industrial Pollution." *Xuehai* 2 (2012): 91–97.

Lu, Jixia, and Anna Lora-Wainwright. "Historicizing Sustainable Livelihoods: A Pathways Approach to Lead Mining in Rural Central China." *World Development* 62 (2014): 189–200.

Mah, Alice. *Industrial Ruination, Community, and Place: Landscapes and Legacies of Urban Decline*. Toronto: University of Toronto Press, 2012.

Martínez-Alier, Joan. "Between Activism and Science: Grassroots Concepts for Sustainability Coined by Environmental Justice Organizations." *Journal of Political Ecology* 21 (2014): 19–60.

Martínez-Alier, Joan. *The Environmentalism of the Poor*. Cheltenham, UK: Edward Elgar, 2003.

Martínez-Alier, Joan, Leah Temper, Daniela Del Bene, and Arnim Scheidel. "Is There a Global Environmental Justice Movement?" *The Journal of Peasant Studies* 43, no. 3 (2016): 731–755.

McAdam, D., J. D. McCarthy, and M. N. Zald, eds. *Comparative Perspectives on Social Movements: Political Opportunities, Mobilizing Structures, and Cultural Framings*. Cambridge: Cambridge University Press, 1996.

McAdam, Doug, and Hilary Schaffer Boudet. *Putting Social Movements in their Place: Explaining Opposition to Energy Projects in the United States, 2000–2005*. Cambridge: Cambridge University Press, 2012.

McGregor, Richard. "750,000 a Year Killed by Chinese Pollution," *Financial Times*, July 2, 2007. Accessed March 15, 2014. http://www.ft.com/cms/s/0/8f40e248-28c7 -11dc-af78-000b5df10621.html.

Mellucci, Alberto. *Nomads of the Present: Social Movements and Individual Needs in Contemporary Society*. Philadelphia, PA: Temple University Press, 1989.

Mertha, Andrew. *China's Water Warriors: Citizen Action and Policy Change*. Ithaca: Cornell University Press, 2010. First published 2008.

Michaels, David. *Doubt is Their Product: How Industry's Assault on Science Threatens Your Health*. London: Oxford University Press, 2008.

Michelson, Ethan. "Climbing the Dispute Pagoda: Grievances and Appeals to the Official Justice System in Rural China." *American Sociological Review* 72 (2007): 459–485.

Ministry of Environmental Protection of the People's Republic of China. "Guard Against and Control Risks Presented by Chemicals to the Environment During the 12th Five-Year Plan (2011–2015)," 1–36. Published January 2013 (2013a). Accessed March 10, 2013. http://www.zhb.gov.cn/gkml/hbb/bwj/201302/W020130220539067366659.pdf.

Ministry of Environmental Protection of the People's Republic of China. "Bulletin on China's Environmental Conditions 2012." Published May 25, 2013 (2013b). Accessed September 9, 2013. http://www.mep.gov.cn/gkml/hbb/qt/201306/W020130606578292022739.pdf.

Ministry of Environmental Protection of the People's Republic of China. "The Ministry of Environmental Protection's 3 Large Campaigns on Solid Waste Pollution Prevention." Published January 7, 2013 (2013c). Accessed January 22, 2016. http://www.mep.gov.cn/zhxx/hjyw/201301/t20130107_244705.htm.

Ministry of Environmental Protection of the People's Republic of China. "Environmental Quality Standards for Soil, GB 156128—1995." Issued July 13, 1995; promulgated March 1, 1996. Accessed September 26, 2016. http://kjs.mep.gov.cn/hjbhbz/bzwb/trhj/trhjzlbz/199603/W020070313485587994018.pdf.

Ministry of Land and Resources of the People's Republic of China. "The Most Promising Lead-Zinc Mine in China Will Bid to Public Next Year and Mining Will Start within Five Years." May 22, 2012. Accessed September 12, 2012. http://www.mlr.gov.cn/xwdt/kyxw/201205/t20120522_1101177.htm.

Ministry of Land and Resources of the People's Republic of China. "Hunan Largest Lead-Zinc Mineral Reserves Found in Xiangxi and the Total Available Resources Reaches 10 Million Tons." November 25, 2004. Accessed September 12, 2012. http://www.mlr.gov.cn/xwdt/jrxw/200411/t20041125_613051.htm.

Minter, Adam. *Junkyard Planet*. New York: Bloomsbury, 2013.

Morton, Katherine. *International Aid and China's Environment: Taming the Yellow Dragon*. Abingdon, UK: Routledge, 2005.

Moskvitch, Katia. "Unused E-Waste Discarded in China Raises Questions." *BBC News*, April 20, 2012. Accessed January 20, 2016. http://www.bbc.com/news/technology -17782718.

Munck, Ronaldo. "The Precariat: A View from the South." *Third World Quarterly* 34, no. 5 (2013): 747–762.

Munro, Neil. "Profiling the Victims: Public Awareness of Pollution-Related Harm in China." *Journal of Contemporary China* 23, no. 86 (2014): 314–329.

Murphy, Michelle. *Sick Building Syndrome and the Problem of Uncertainty: Environmental Politics, Technoscience, and Women Workers*. Durham, NC: Duke University Press, 2006.

Murphy, Rachel, ed. *Labour Migration and Social Development in Contemporary China*. Abingdon, UK: Routledge, 2009.

Murphy, Rachel. *How Migrant Labour Is Changing Rural China*. Cambridge: Cambridge University Press, 2002.

*Nanfang Dushi*. "A Diary of Three Villages of Death," November 5, 2007 (2007a). Accessed November 21, 2007. http://www.nddaily.com/A/html/2007-11/05/content _299441.htm.

*Nanfang Dushi*. "China's Water Crisis," November 2, 2007 (2007b). Accessed November 21, 2007. http://www.nddaily.com/sszt/watercrisis/.

Nash, June. *We Eat the Mines and the Mines Eat Us*. New York: Columbia University Press, 1979.

Nixon, Rob. *Slow Violence and the Environmentalism of the Poor*. Cambridge, MA: Harvard University Press, 2011.

O'Brien, Kevin. "Rightful Resistance Revisited." Special issue, *Journal of Peasant Studies* 40, no. 6 (2013): 1051–1062.

O'Brien, Kevin, and Lianjiang Li. *Rightful Resistance in Rural China*. Cambridge: Cambridge University Press, 2006.

Oi, Jean C. *Rural China Takes Off: Institutional Foundations of Economic Reform*. Berkeley: University of California Press, 1999.

Olson, Valerie. "The Ecobiopolitics of Space Biomedicine." *Medical Anthropology* 29, no. 2 (2010): 170–193.

Ottinger, Gwen, and Benjamin R. Cohen. *Technoscience and Environmental Justice: Expert Cultures in a Grassroots Movement*. Cambridge, MA: MIT Press, 2011.

Ottinger, Gwen. *Refining Expertise: How Responsible Engineers Subvert Environmental Justice Challenges*. New York: New York University Press, 2013.

Peet, Richard, and Michael Watts. *Liberation Ecologies: Environment, Development, Social Movements*. 2nd ed. London: Routledge, 2004.

Pellow, David. *Resisting Global Toxics: Transnational Movements for Environmental Justice*. Cambridge, MA: MIT Press, 2007.

*People's Daily*. "Opinions of the Central Committee of the Communist Party of China and the State Council on Further Promoting the Development of Ecological Civilization." Central Document No. 12. Issued April 25, 2015. Published June 6, 2015. Accessed January 22, 2016. http://paper.people.com.cn/rmrb/html/2015-05/06/nw.D110000renmrb_20150506_3-01.htm.

Petryna, Adriana. *Life Exposed: Biological Citizens after Chernobyl*. Princeton, NJ: Princeton University Press, 2002.

Phillips, Tarryn. "Repressive Authenticity in the Quest for Legitimacy: Surveillance and the Contested Illness Lawsuit." *Social Science & Medicine* 75 (2012): 1762–1768.

Proctor, Robert. *Cancer Wars: How Politics Shapes What We Know and Don't Know About Cancer*. New York: Basic Books, 1995.

Proctor, R., and L. Schiebinger, eds. *Agnotology: The Making and Unmaking of Ignorance*. Palo Alto, CA: Stanford University Press, 2008.

Pun, Ngai. *Made in China: Women Factory Workers in a Global Workplace*. Durham, NC: Duke University Press, 2005.

Pun, Ngai, and Jenny Chan. "The Spatial Politics of Labor in China: Life, Labor, and a New Generation of Migrant Workers." *South Atlantic Quarterly* 112, no. 1 (2013): 179–190.

Pun, Ngai, and Huilin Lu. "Unfinished Proletarianization: Self, Anger, and Class Action among the Second Generation of Peasant-Workers in Present-Day China." *Modern China* 36, no. 5 (2010): 493–519.

Ran, S. "Land Use and Livelihood Change." Paper presented at the SSRC FORHEAD Annual Conference, Beijing, China, November 7–9, 2012.

Robbins, Paul. *Political Ecology: A Critical Introduction*. Oxford: Blackwell, 2004.

Rofel, Lisa. *Other Modernities: Gendered Yearnings in China after Socialism*. Berkeley: University of California Press, 1999.

Rucevska, Ieva, Christian Nellemann, Nancy Isarin, Wanhua Yang, Liu Ning, Keili Yu, Siv Sandnæs, et al. "Waste Crime—Waste Risks: Gaps in Meeting the Global Waste Challenge (A UNEP Rapid Response Assessment)." United Nations Environment Programme and GRID-Arendal, May 12, 2015. Accessed January 20, 2016. http://www.unep.org/environmentalgovernance/Portals/8/documents/rra-wastecrime.pdf.

Saunders, Clare. *Environmental Networks and Social Movement Theory*. London: Bloomsbury, 2013.

Saxton, Dvera. "Strawberry Fields as Extreme Environments: The Ecobiopolitics of Farmworker Health." *Medical Anthropology* 34, no. 2 (2014): 166–183.

Schlosberg, David. *Defining Environmental Justice: Theories, Movements and Nature*. Oxford: Oxford University Press, 2007.

Schulz, Yvan. "Towards a New Waste Regime? Critical Reflections on China's Shifting Market for High-Tech Discards." *China Perspectives* 3 (2015): 43–50.

Scott, James. *Weapons of the Weak: Everyday Forms of Peasant Resistance*. New Haven, CT: Yale University Press, 1985.

Sen, Amartya. *The Idea of Justice*. London: Allen Lane, 2009.

Shantou City Environmental Protection Bureau Net. "Municipal Environmental Protection Bureau's 8 Big Measures to Implement the Spirit of the Provincial Government Meeting on Guiyu Pollution Clean-up Work." Published May 23, 2013. Accessed September 9, 2013. http://www.stepb.gov.cn/zwxx/wshbdt/201305/t20130523_6551.html.

Shapiro, Judith. *China's Environmental Challenges*. Cambridge: Polity Press, 2012.

Shapiro, Judith. *Mao's War against Nature: Politics and the Environment in Revolutionary China*. Cambridge: Cambridge University Press, 2001.

Shevory, Thomas. *Toxic Burn: The Grassroots Struggle against the WTI Incinerator*. Minneapolis, MN: Minnesota University Press, 2007.

Silbergeld, Ellen K. *The Elimination of Lead from Gasoline: Impacts of Lead in Gasoline on Human Health, and the Costs and Benefits of Eliminating Lead Additives*. Washington, DC: The World Bank, 1996.

Silbergeld, Ellen K. "The International Dimensions of Lead Exposure." *International Journal of Occupational and Environmental Health* 1, no. 4 (1995): 336–348.

Singer, Merrill. "Down Cancer Alley: The Lived Experience of Health and Environmental Suffering in Louisiana's Chemical Corridor." *Medical Anthropology Quarterly* 25, no. 2 (2011): 141–163.

Small, Mario Luis. "How Many Cases Do I Need? On Science and the Logic of Case Selection in Field-Based Research." *Ethnography* 10, no. 1 (2009): 5–38.

Smartt Gullion, Jessica. *Fracking the Neighborhood: Reluctant Activists and Natural Gas Drilling*. Cambridge, MA: MIT Press, 2015.

Snow, David, and Robert Benford. "Master Frames and Cycles of Protest." In *Frontiers in Social Movement Theory*, edited by A. Morris and C. Mueller, 133–155. New Haven, CT: Yale University Press, 1992.

Solinger, Dorothy. *Contesting Citizenship in Urban China: Peasant Migrants, the State, and the Logic of the Market*. Berkeley: University of California Press, 1999.

*South China Morning Post*. "Ningbo Protests against Growth at any Cost." Editorial, October 30, 2012. Accessed January 22, 2016. http://www.scmp.com/comment/insight-opinion/article/1072726/ningbo-protests-against-growth-any-cost.

Spires, Anthony. "Contingent Symbiosis and Civil Society in an Authoritarian State: Understanding the Survival of China's Grassroots NGOs." *American Journal of Sociology* 117 (2011): 1–45.

Standing, Guy. *The Precariat: The New Dangerous Class*. London: Bloomsbury, 2014.

State Council Information Office of the People's Republic of China. *White Book of Poverty Alleviation in Rural China*. Beijing: The State Council Leading Group Office of Poverty Alleviation and Development, 2002.

Steingraber, Sandra. *Living Downstream: An Ecologist's Personal Investigation of Cancer and the Environment*. 2nd ed. Philadelphia, PA: Da Capo Press, 2010.

Steinhardt, H. Christoph, and Fengshi Wu. "In the Name of the Public: Environmental Protest and the Changing Landscape of Popular Contention in China." *China Journal (Canberra, A.C.T.)* 75 (2015): 61–82.

Stern, Rachel. *Environmental Litigation in China: A Study in Political Ambivalence*. Cambridge: Cambridge University Press, 2013.

Stern, Rachel. "From Dispute to Decision: Suing Polluters in China." *China Quarterly* 206 (2011): 294–312.

Su, Xiaokang, and Perry Link. "A Collapsing Natural Environment?" In *Restless China*, edited by Perry Link, Richard P. Madsen, and Paul G. Pickowicz, 213–233. Lanham, UK: Rowman and Littlefield, 2013.

Swartz, David. *Culture and Power: The Sociology of Pierre Bourdieu*. Chicago, IL: University of Chicago Press, 1997.

Szasz, Andrew. *Shopping Our Way to Safety. How We Changed from Protecting the Environment to Protecting Ourselves*. Minneapolis: Minnesota University Press, 2007.

Szasz, Andrew. *Ecopopulism: Toxic Waste and the Movement for Environmental Justice*. Minneapolis, MN: University of Minnesota Press, 1994.

Tang, Hao. "Shifang: A Crisis of Local Rule." *China Dialogue*, July 18, 2012. Accessed January 21, 2016. https://www.chinadialogue.net/article/show/single/en/5049-Shifang-a-crisis-of-local-rule.

Tarrow, Sidney. "Struggling to Reform: Social Movements and Policy Change during Cycles of Protest." Cornell University Western Societies Program: Occasional Paper No. 15. Ithaca, NY: Cornell University Press, 1983.

Tejada, Carlos. "China Move Reflects Sensitivity on Pollution" *Wall Street Journal*, July 30, 2012. Accessed September 26, 2016. http://online.wsj.com/article/ SB10000872396390444130304577556942080632120.html.

Tesh, Sylvia. *Uncertain Hazards: Environmental Activists and Scientific Proof.* Ithaca, NY: Cornell University Press, 2000.

Tilly, Charles, and Sydney Tarrow. *Contentious Politics.* Boulder, CO: Paradigm Publishers, 2007.

Tilt, Bryan. "Industrial Pollution and Environmental Health in Rural China: Risk, Uncertainty and Individualization." *China Quarterly* 214 (2013): 283–301.

Tilt, Bryan. *The Struggle for Sustainability in Rural China.* New York: Columbia University Press, 2010.

Tilt, Bryan. "Perceptions of Risk from Industrial Pollution in China: A Comparison of Occupational Groups." *Human Organization* 65, no. 2 (2006): 115–127.

Tilt, Bryan, and Qing Xiao. "Media Coverage of Environmental Pollution in the People's Republic of China: Responsibility, Cover-up and State Control." *Media Culture & Society* 32, no. 2 (2010): 225–245.

Tong, Shilu, Peter A. Baghurst, Michael G. Sawyer, Jane Burns, and Anthony J. McMichael. "Declining Blood Lead Levels and Changes in Cognitive Function during Childhood: The Port Pirie Cohort Study." *Journal of the American Medical Association* 280, no. 22 (1998): 1915–1919.

Tong, Xin, and Jici Wang. "The Shadow of the Global Network: E-Waste Flows to China." In *Economies of Recycling: The Global Transformation of Materials, Values and Social Relations*, edited by Catherine Alexander and Joshua Reno, 98–117. London: Zed Books, 2012.

Tong, Xin, and Lin Yan. "From Legal Transplants to Sustainable Transition: Extended Producer Responsibility in Chinese Waste Electrical and Electronic Equipment Management." *Journal of Industrial Ecology* 17, no. 2 (2013): 199–212.

Tong, Xin, Jingyan Li, Dongyan Tao, and Yifan Cai. "Re-Making Spaces of Conversion: Deconstructing Discourses of E-Waste Recycling in China." *Area* 47, no. 1 (2015): 31–39.

*Tuoitrenews.* "UN Report Reveals Major Criminal Activities in East Asia-Pacific." August 15, 2013. Accessed January 20, 2016. http://tuoitrenews.vn/society/12236/ un-report-reveals-major-criminal-activities-in-east-asiapacific.

*Under the Dome.* Online documentary film. Produced and directed by Chai Jing, 2015. Accessed January 19, 2016. https://www.youtube.com/watch?v=T6X2uwlQGQM.

Van Rooij, Benjamin. "The People vs. Pollution: Understanding Citizen Action against Pollution in China." *Journal of Contemporary China* 19, no. 63 (2010): 55–77.

Van Rooij, Benjamin. *Regulating Land and Pollution in China: Lawmaking, Compliance, and Enforcement: Theory and Cases*. Leiden: Leiden University Press, 2006.

Van Rooij, Benjamin, Anna Lora-Wainwright, Yunmei Wu, and Yiyun Zhang. "Activist Acquiescence: Power, Pollution, and Access to Justice in a Chinese Village." UC Irvine School of Law, Research Paper No. 2014–22, March 7, 2014. Accessed January 22, 2016. http://papers.ssrn.com/sol3/papers.cfm?abstract_id=2406219.

Van Rooij, Benjamin, Anna Lora-Wainwright, Yunmei Wu, and Yiyun Zhang. "The Compensation Trap: The Limits of Community-Based Pollution Regulation in China." *Pace Environmental Law Review* 29, no. 3 (2012): 701–745.

Vanderlinden, Lisa. "Left in the Dust: Negotiating Environmental Illness in the Aftermath of 9/11." *Medical Anthropology* 30, no. 1 (2011): 30–55.

Waldman, Linda. *The Politics of Asbestos: Understandings of Risk, Disease and Protest*. London: Earthscan, 2011.

Waldman, Linda. "'Show Me the Evidence': Mobilisation, Citizenship, and Risk in Indian Asbestos Issues." IDS Working Papers 329, pp. 1–48, Institute of Development Studies, Brighton, UK, 2009.

Walker, Gordon. *Environmental Justice: Concepts, Evidence and Politics*. London: Routledge, 2012.

Wang, Alex. "New Documentary Follows the Struggle to Save China's Environment," *Huffington Post*, January 13, 2011. Accessed January 20, 2016. http://www.huffingtonpost.com/alex-wang/the-warriors-of-qiugang-a_b_808382.html.

Wang, Alex. "The Role of Law in Environmental Protection in China: Recent Developments." *Journal of Environmental Law* 195 (2006): 195–223.

Wang, Feng, Ruediger Kuehr, Daniel Ahlquist, and Jinhui Li. "E-Waste in China: A Country Report." United Nations University—Institute for Sustainability and Peace. StEPs Green Paper Series, April 5, 2013. Accessed January 20, 2016. http://ewasteguide.info/files/Wang_2013_StEP.pdf.

Wang, Xiaolin. "Integrating Rural Access to Safe Drinking Water into Public Planning and Budgeting." September 30, 2013. FORHEAD funded projects. Accessed June 9, 2016. http://www.forhead.org/en/support/ssrc/zs/2013/0607/940.html.

Wang, Yan. "Death Maps," *News China Magazine*, October 2013 (2013a). Accessed November 1, 2013. http://old.newschinamag.com/magazine/death-maps.

Wang, Yan. "The Voice of the Polluted," *News China Magazine*, September 2013 (2013b). Accessed November 1, 2013. http://old.newschinamag.com/magazine/the-voice-of-the-polluted.

*Warriors of Qiugang, The*. Documentary film. Directed by Ruby Yang. Thomas Lennon Films & Chang Ai Media Project, 2010.

Watson, Ivan. "China: The Electronic Wastebasket of the World." *CNN*, May 31, 2013. Accessed January 20, 2016. http://edition.cnn.com/2013/05/30/world/asia/china-electronic-waste-e-waste/index.html.

Weller, Robert. *Discovering Nature: Globalization and Environmental Culture in China and Taiwan*. Cambridge: Cambridge University Press, 2006.

Wells-Dang, Andrew. *Civil Society Networks in China and Vietnam: Informal Pathbreakers in Health and the Environment*. Basingstoke, UK: Palgrave Macmillan, 2012.

Wertime, David. "Chinese State Media Shares Powerful Map of 'Cancer Villages' Creeping Inland," *Tea Leaf Nation*, February 22, 2013. Accessed January 22, 2016. http://chinafile.luxcerclients.com/china-s-state-run-media-shares-powerful-map-cancer-villages-creeping-inland.

Whiting, Susan H. "Fiscal Reform and Land Public Finance: Zouping County in National Context." In *China's Local Public Finance in Transition*, edited by Joyce Yanyun Man and Yu-Hung Hong, 125–143. Cambridge, MA: Lincoln Institute of Land Policy, 2010.

Whiting, Susan H. *Power and Wealth in Rural China: The Political Economy of Institutional Change*. Cambridge: Cambridge University Press, 2001.

WHO (World Health Organization). "Inorganic Lead." *Environmental Health Criteria 165, International Programme on Chemical Safety*. 1995. Accessed October 26, 2016. http://www.inchem.org/documents/ehc/ehc/ehc165.htm.

Williams, Christopher. "An Environmental Victimology." In *Environmental Victims: New Risks, New Injustice*, edited by Christopher Williams, 3–26. London: Earthscan, 1998.

Williams, Glynn, and Emma Mawdsley. "Postcolonial Environmental Justice: Government and Governance in India." *Geoforum* 37, no. 5 (2006): 660–670.

Williams, Raymond. *Marxism and Literature*. Oxford: Oxford University Press, 1977.

World Bank and State Environmental Protection Administration (SEPA). *The Cost of Pollution in China: Economic Estimates of Physical Damages*. February 1, 2007. Accessed January 20, 2012. http://siteresources.worldbank.org/INTEAPREGTOPENVIRONMENT/Resources/China_Cost_of_Pollution.pdf.

Wright, Tim. *The Political Economy of the Chinese Coal Industry: Black Gold and Blood-Stained Coal*. London: Routledge, 2011.

Wu, Fengshi. "Environmental Activism in Provincial China." *Journal of Environmental Policy and Planning* 15, no. 1 (2013): 89–108.

Wynne, Brian. "May the Sheep Safely Graze? A Reflexive View of the Expert/Lay Knowledge Divide." In *Risk, Environment and Modernity: Towards a New Ecology*, edited by Scott Lash, Bronislaw Szeszynski, and Brian Wynne, 44–84. London: Sage, 1996.

Xinhuanet. "234 People Died in the Shanxi Dam Break Accident and the Provincial Governor Take the Blame and Resign." September 15, 2008. Accessed January 1, 2013. http://news.jinghua.cn/351/c/200809/15/n1777651.html.

Yan, Hairong. *New Masters, New Servants. Migration, Development and Women Workers in China*. Durham, NC: Duke University Press, 2008.

Yan, Yunxiang. *The Individualisation of Chinese Society*. Oxford: Berg, 2009.

Yan, Yunxiang. *Private Life Under Socialism: Love, Intimacy, and Family Change in a Chinese Village, 1949–1999*. Stanford, CA: Stanford University Press, 2003.

Yang, Chuanming, and Qianhua Fang. "A Village of Death and Its Hopes for the Future," *Nanfang News Evening Edition*, November 18, 2005. Accessed March 10, 2007. http://www.southcn.com/news/dishi/shaoguan/ttxw/200511180238.htm.

Yang, Gonghuan, and Dafang Zhuang. *Atlas of the Water Environment and Digestive Cancer Mortality in the Huai River Basin*. Beijing: Zhongguo Ditu Chubanshe, 2013.

Yang, Guobin. "Contesting Food Safety in Chinese Media: Between Hegemony and Counter-Hegemony." Special issue, *China Quarterly* 214 (2013): 337–355.

Yang, Guobin. *The Power of the Internet in China: Citizen Activism Online*. New York: Columbia University Press, 2011. (Previously published in 2009).

Yang, Guobin. "Brokering Environment and Health in China: Issue Entrepreneurs of the Public Sphere." Special issue, *Journal of Contemporary China* 19, no. 63 (2010a): 101–118.

Yang, Guobin. "Civic Environmentalism." In *Reclaiming Chinese Society. The New Social Activism*, edited by You-tien Hsing and Ching Kwan Lee, 119–139. London: Routledge, 2010b.

Yang, Guobin. "Environmental NGOs and Institutional Dynamics in China." *China Quarterly* 181 (2005): 46–66.

Yang, Guobin, and Craig Calhoun. "Media, Civil Society and the Rise of a Green Public Sphere in China." In *China's Embedded Activism: Opportunities and Constraints of a Social Movement*, edited by Peter Ho and Richard Edmonds, 69–88. London: Routledge, 2008. Originally published in *China Information* 21, no. 2 (2007): 211–236.

Yip, Winnie Chi-Man, William C. Hsiao, Wen Chen, Shanlian Hu, Jin Ma, and Alan Maynard. "Early Appraisal of China's Huge and Complex Health-Care Reforms." *Lancet* 379, no. 9818 (2012): 833–842.

Zhang, Amy. "Rational Resistance: Homeowner Contentions against Waste Incineration in Guangzhou." Special feature, *China Perspectives* 2 (2014): 45–52.

Zhang, Joy Y., and Michael Barr. *Green Politics in China: Environmental Governance and State-Society Relations*. London: Pluto Press, 2013.

Zhang, Li. *Strangers in the City: Reconfigurations of Space, Power, and Social Networks within China's Floating Population*. Stanford, CA: Stanford University Press, 2001.

Zhang, Liping. "From Guiyu to a Nationwide Policy: E-Waste Management in China." *Environmental Politics* 18, no. 6 (2009): 981–987.

Zhang, X. "Health Risk Identification and Assessment Study for the Residents of Rural Mining Village of Fenghuang County, Western Hunan Province." PhD diss., Chinese Academy of Sciences, 2011.

Zhang, X., L. Yang, Y. Li, H. Li, W. Wang, and Q. Ge. "Estimation of Lead and Zinc Emissions from Mineral Exploitation Based on Characteristics of Lead/Zinc Deposits in China." *Transactions of Nonferrous Metals Society of China* 21 (2011): 2513–2519.

Zhang, X., L. Yang, Y. Li, H. Li, W. Wang, and B. Ye. "Impacts of Lead/Zinc Mining and Smelting on the Environment and Human Health in China." *Environmental Monitoring and Assessment* 184 (2012): 2261–2273.

Zhou, Ying. "The State of Precarious Work in China." *American Behavioral Scientist* 57, no. 3 (2013): 354–372.

Zhu, Haifeng. "Mining Waste Water Harms a Village," *Chinese Mining Newspaper*, January 20, 2001.

Zimring, C., and W. Rathje, eds. *Encyclopaedia of Consumption and Waste: The Social Science of Garbage*. Thousand Oaks, CA: Sage, 2012.

# Index

## Urban and Industrial Environments

Series editor: Robert Gottlieb, Henry R. Luce Professor of Urban and Environmental Policy, Occidental College

Maureen Smith, *The U.S. Paper Industry and Sustainable Production: An Argument for Restructuring*

Keith Pezzoli, *Human Settlements and Planning for Ecological Sustainability: The Case of Mexico City*

Sarah Hammond Creighton, *Greening the Ivory Tower: Improving the Environmental Track Record of Universities, Colleges, and Other Institutions*

Jan Mazurek, *Making Microchips: Policy, Globalization, and Economic Restructuring in the Semiconductor Industry*

William A. Shutkin, *The Land That Could Be: Environmentalism and Democracy in the Twenty-First Century*

Richard Hofrichter, ed., *Reclaiming the Environmental Debate: The Politics of Health in a Toxic Culture*

Robert Gottlieb, *Environmentalism Unbound: Exploring New Pathways for Change*

Kenneth Geiser, *Materials Matter: Toward a Sustainable Materials Policy*

Thomas D. Beamish, *Silent Spill: The Organization of an Industrial Crisis*

Matthew Gandy, *Concrete and Clay: Reworking Nature in New York City*

David Naguib Pellow, *Garbage Wars: The Struggle for Environmental Justice in Chicago*

Julian Agyeman, Robert D. Bullard, and Bob Evans, eds., *Just Sustainabilities: Development in an Unequal World*

Barbara L. Allen, *Uneasy Alchemy: Citizens and Experts in Louisiana's Chemical Corridor Disputes*

Dara O'Rourke, *Community-Driven Regulation: Balancing Development and the Environment in Vietnam*

Brian K. Obach, *Labor and the Environmental Movement: The Quest for Common Ground*

Peggy F. Barlett and Geoffrey W. Chase, eds., *Sustainability on Campus: Stories and Strategies for Change*

Steve Lerner, *Diamond: A Struggle for Environmental Justice in Louisiana's Chemical Corridor*

Jason Corburn, *Street Science: Community Knowledge and Environmental Health Justice*

Peggy F. Barlett, ed., *Urban Place: Reconnecting with the Natural World*

David Naguib Pellow and Robert J. Brulle, eds., *Power, Justice, and the Environment: A Critical Appraisal of the Environmental Justice Movement*

Eran Ben-Joseph, *The Code of the City: Standards and the Hidden Language of Place Making*

Nancy J. Myers and Carolyn Raffensperger, eds., *Precautionary Tools for Reshaping Environmental Policy*

Kelly Sims Gallagher, *China Shifts Gears: Automakers, Oil, Pollution, and Development*

Kerry H. Whiteside, *Precautionary Politics: Principle and Practice in Confronting Environmental Risk*

Ronald Sandler and Phaedra C. Pezzullo, eds., *Environmental Justice and Environmentalism: The Social Justice Challenge to the Environmental Movement*

Julie Sze, *Noxious New York: The Racial Politics of Urban Health and Environmental Justice*

Robert D. Bullard, ed., *Growing Smarter: Achieving Livable Communities, Environmental Justice, and Regional Equity*

Ann Rappaport and Sarah Hammond Creighton, *Degrees That Matter: Climate Change and the University*

Michael Egan, *Barry Commoner and the Science of Survival: The Remaking of American Environmentalism*

David J. Hess, *Alternative Pathways in Science and Industry: Activism, Innovation, and the Environment in an Era of Globalization*

Peter F. Cannavò, *The Working Landscape: Founding, Preservation, and the Politics of Place*

Paul Stanton Kibel, ed., *Rivertown: Rethinking Urban Rivers*

Kevin P. Gallagher and Lyuba Zarsky, *The Enclave Economy: Foreign Investment and Sustainable Development in Mexico's Silicon Valley*

David N. Pellow, *Resisting Global Toxics: Transnational Movements for Environmental Justice*

Robert Gottlieb, *Reinventing Los Angeles: Nature and Community in the Global City*

David V. Carruthers, ed., *Environmental Justice in Latin America: Problems, Promise, and Practice*

Tom Angotti, *New York for Sale: Community Planning Confronts Global Real Estate*

Paloma Pavel, ed., *Breakthrough Communities: Sustainability and Justice in the Next American Metropolis*

Anastasia Loukaitou-Sideris and Renia Ehrenfeucht, *Sidewalks: Conflict and Negotiation over Public Space*

David J. Hess, *Localist Movements in a Global Economy: Sustainability, Justice, and Urban Development in the United States*

Julian Agyeman and Yelena Ogneva-Himmelberger, eds., *Environmental Justice and Sustainability in the Former Soviet Union*

Jason Corburn, *Toward the Healthy City: People, Places, and the Politics of Urban Planning*

JoAnn Carmin and Julian Agyeman, eds., *Environmental Inequalities Beyond Borders: Local Perspectives on Global Injustices*

Louise Mozingo, *Pastoral Capitalism: A History of Suburban Corporate Landscapes*

Gwen Ottinger and Benjamin Cohen, eds., *Technoscience and Environmental Justice: Expert Cultures in a Grassroots Movement*

Samantha MacBride, *Recycling Reconsidered: The Present Failure and Future Promise of Environmental Action in the United States*

Andrew Karvonen, *Politics of Urban Runoff: Nature, Technology, and the Sustainable City*

Daniel Schneider, *Hybrid Nature: Sewage Treatment and the Contradictions of the Industrial Ecosystem*

Catherine Tumber, *Small, Gritty, and Green: The Promise of America's Smaller Industrial Cities in a Low-Carbon World*

Sam Bass Warner and Andrew H. Whittemore, *American Urban Form: A Representative History*

John Pucher and Ralph Buehler, eds., *City Cycling*

Stephanie Foote and Elizabeth Mazzolini, eds., *Histories of the Dustheap: Waste, Material Cultures, Social Justice*

David J. Hess, *Good Green Jobs in a Global Economy: Making and Keeping New Industries in the United States*

Joseph F. C. DiMento and Clifford Ellis, *Changing Lanes: Visions and Histories of Urban Freeways*

Joanna Robinson, *Contested Water: The Struggle Against Water Privatization in the United States and Canada*

William B. Meyer, *The Environmental Advantages of Cities: Countering Commonsense Antiurbanism*

Rebecca L. Henn and Andrew J. Hoffman, eds., *Constructing Green: The Social Structures of Sustainability*

Peggy F. Barlett and Geoffrey W. Chase, eds., *Sustainability in Higher Education: Stories and Strategies for Transformation*

Isabelle Anguelovski, *Neighborhood as Refuge: Community Reconstruction, Place-Remaking, and Environmental Justice in the City*

Kelly Sims Gallagher, *The Global Diffusion of Clean Energy Technology: Lessons from China*

Vinit Mukhija and Anastasia Loukaitou-Sideris, eds., *The Informal American City: Beyond Taco Trucks and Day Labor*

Roxanne Warren, *Rail and the City: Shrinking Our Carbon Footprint and Reimagining Urban Space*

Marianne Krasny and Keith Tidball, *Civic Ecology: Adaptation and Transformation from the Ground Up*

Erik Swyngedouw, *Liquid Power: Contested Hydro-Modernities in Twentieth-Century Spain*

Duncan McLaren and Julian Agyeman, *Sharing Cities: A Case for Truly Smart and Sustainable Cities*

Jessica Smartt Gullion, *Fracking the Neighborhood: Reluctant Activists and Natural Gas Drilling*

Nicholas Phelps, *Sequel to Suburbia: Glimpses of America's Post-Suburban Future*

Shannon Elizabeth Bell, *Fighting King Coal: The Challenges to Micromobilization in Central Appalachia*

Theresa Enright, *The Making of Grand Paris: Metropolitan Urbanism in the Twenty-First Century*

Anna Lora-Wainwright, *Resigned Activism: Living with Pollution in Rural China*